POLAROGRAPHIC METHODS IN ANALYTICAL CHEMISTRY

POLAROGRAPHIC METHODS IN ANALYTICAL CHEMISTRY

M. G. ARORA

ANMOL PUBLICATIONS PVT LTD
NEW DELHI-110 002 (INDIA)

ANMOL PUBLICATIONS PVT LTD
4374/4B, Ansari Road, Daryaganj
New Delhi-110 002

Polarographic Methods in Analytical Chemistry

First Edition 1996

ISBN 81-7041-917-4

PRINTED IN INDIA

Published by J. L. Kumar for Anmol Publications Pvt Ltd, New Delhi and Printed at Mehra Offset Press, Delhi.

Preface

This book is an attempt to make available to the reader various polarographic methods in a comprehensive and easy to understand form. The book deals in detail with the apparatus used in polarographic methods and explains the use of all such apparatus. Various currents and different types of systems to be studied in undergraduate/postgraduate classes have been discussed in details.

It is hoped that the book will be quite useful to teachers, research scholars and even undergraduate students. Applications of polarography may be of interest of general reader as well.

For convenience to the reader the book has been divided in 12 chapters. Each chapter is complete in itself.

Suggestions for any further improvement in the book shall be gratefully accepted.

Author

Contents

1

Introduction to Polarography

Basic Concepts in Polarography

Electrochemistry in the relationship between electrical properties and chemical reaction. As applied in the area of analytical chemistry this usually involves the measurement of some electrical property under conditions which, directly or indirectly, permit an association between the magnitude of property measured and the concentration of some particular chemical species. Such measurements are nearly always made in solution environments.

The electrical properties most commonly measured involve either voltage, current or resistance or combination of these.

The polarographic method is probably the most widely used Faradaic electrochemical technique. It is one of the methods in which transfer of electrons takes place between an electrode and the ions, atoms or molecules present in the surrounding media. The polarographic method employs a Dropping Mercury Electrode (DME) during electrolysis.

Polarography as an analytical technique was conceived by J. Heyrowsky who desired the basic technique in a major series of papers starting about 1922. The technique received fresh impeters in 1935 from the mathematics proposed by II kovic to describe the poloargraphic current.

Heyrowsky and his Japanese coworker Shikata fabricated an instrument known as *polarography*, Polarograph can record the electrolytic curves automatically.

The plot of current against applied voltage for a sample solution is known as a *polarogram.*

Voltammetry and *Polarography* are based an measurement of the diffusion controlled limiting current between an indicator electrode and a refrence electrode where a fixed potential is maintained between the electrodes.

Voltammetry is the study of the entire course of the applied potential Vs current curve as this exists for an electrode of very small area or microelectrode and a refrence electrode of relatively large area. Where such a study involves a DME as a microelectrode, this particular aspect of voltammetry is called *polarography*. Where the principles of *Voltammetry* and *polarography* are applied to plot the current Vs volume of titrant relationship during a titration to locate and point the technique is known as amperometric titrations.

Since the area of microelectrode used in these applications is extremely small, any current flowing through the solution will also be extremely small in the microampere range.

Theory of Polarography

The theory of polarography is rather complicated. Briefly the method depends an concentration polarisation of the DME and measurement of diffusion current. A metal ion in solution will not be deposited on the cathode as metal and no current will flow until the potential between the solution and cathode reaches a certain critical value. As the potential increases above this value the rate of reduction and the flow of current increase until a thin layer of solution around the cathode has been depleted of the ion being reduced. The current is then limited by the rate of diffusion of reducible ions through this layer. The rate of diffusion depends upon the concentration and diffusion constant of the ion in the solution, the potential difference between the boundries of the diffusion layer and the viscosity of the solutions. If a solution contains a large concentration of other ions (usually supplied by the addition of supporting electrolyte), the potential difference across the diffusion layer is negligible and the rate of diffusion remains constant as the potential is increased. The limiting current, known as diffusion current, is also independent of increase in potential. The viscosity can be kept constant by keeping the temperature and composition of the solution constant, the diffusion current is then proportional to the concertration of the ion in solution and hence the amount of ion present in a solution can be determined by comparing the diffusion currents in the solution and in a standard solution by known concentration. The diffusion current is simply measured from the height of the step in current applied voltage curve. The voltage at which half the diffusion current is reached is characteristic of the particular ion concerned and is referred to as the half-wave potential. If their half-wave potentials are not too close a series of metals may be determined in the same solution.

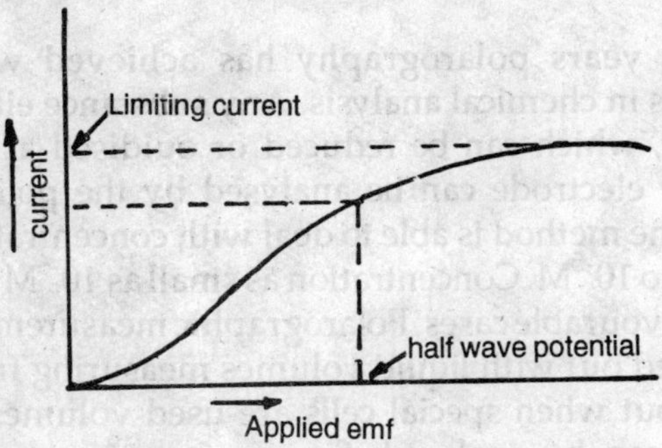

Fig. 1.1. Current Vs applied emf in electrolysis.

The current-potential curve is known as a *polarography* or *polarogram.*

When the solution contains more than one reducible ions, the polarogram takes the shape as shown in Fig. 1.2. Every ion gives characteristic wave unit with definite wave potential. The ion with lowest discharge potential would be discharged first. The nature of the ion in established by the half-wave potential and their quantities by the height of respective waves.

Iristead of DME, some solid electrodes made from metals and some other materials like glassy carbon and graphite have also been used for polarographic studies. Such electrodes have been found useful for carrying out polarographic studies. Outside the potential range of DME, specially at positive potentials. Special electrodes have been made for analysis of materials at micro-and sub-micro levels employing derived polarographic techniques like Differential Pulse Polarography (DPP).

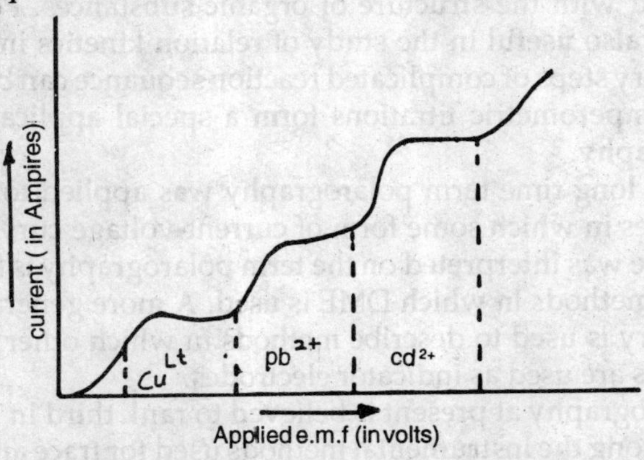

Fig. 1.2. A Polarogram for multi-compound system.

During recent years polarography has achieved wide spread applications in chemical analysis. Any substance either molecular or ionic, which can be reduced or ouidised at the dropping mercury electrode can be analysed by the polarographic method. The method is able to deal with concentration in the range 10^{-2}M to 10^{-5}M. Concentration as small as 10^{-6}M can be determined in favourable cases. Polarographic measurement are generally carried out with liquid volumes measuring from 0.5 ml to 20.0 ml but when special cells are used volumes as small as 0.01 ml can be analysed. Since microcurrents are measured and the amount of chemical change in solution is negligible determinations can be operated on the same sample a number of times.

Along with the growth and increased use of conventional polarography additional techniques as Differential polarography, Derivative polarograp, A.C. polarography, Tast polarography, Square wave polarography, Oscillographic polarography (using single sweep and multiple sweep methods), constant current polarography etc., have been developed. These methods have further enlarged the scope of polarography as an analytical tool. Using some of these techniques analytical applications of polarography can be extended into the range 10^{-9}M, obtaining in most cases a clearer resolution of waves which are too near to be resolved by the conventional polarographic techniques.

Polarographic method has assumed importance in the determination of physico-chemical parameters which can be correlated with the structure of organic substances. Polarography in also useful in the study of relation kinetics in which elementary steps of complicated reaction sequance can be identified. Amperometric titrations form a special application of polarography.

For a long time term polarography was applied to all the techniques in which some form of current-voltage curve or its derivative was interpreted on the term polarography is limited to those methods in which DME is used. A more general term voltametry is used to describe methods in which other microelectrodes are used as indicator electrodes.

Polarography at present is believed to rank third in importance, among the instrumental methods used for trace analysis, being superceded only by mass spectroscopy and spectrophotometry.

2

Polarographic Apparatus

As discussed in previous chapter the limiting diffusion current is a function of the solution concentration, and provided that the diffusion coefficient and diffusion layer thickness are constant values, the dependence of solution concentration on the limiting current can be represented as

$$C_o = k.i_{lim.}$$

where *k* is a constant.

In other words, after measuring the limiting diffusion current, one can calculate the concentration of ions in the solution. This dependence underlies the polarographic method of analysis proposed by the czech scientist J. Heirovsky.

The set up used in the polarographic method is shown schematically in Fig. 2.1. The set up employs a dropping mercury electrode (DME) in the form of a glass capillary tube from which the mercury drops come out. The advantage of dropping mercury electrode is that its surface is continuously renewed during measurements. The mercury drops at the capillary tube end act as the cathode and the mercury at the bottom of the vessel is anode.

The basic apparatus for the polarographic analysis is shown in Fig. 2.1.

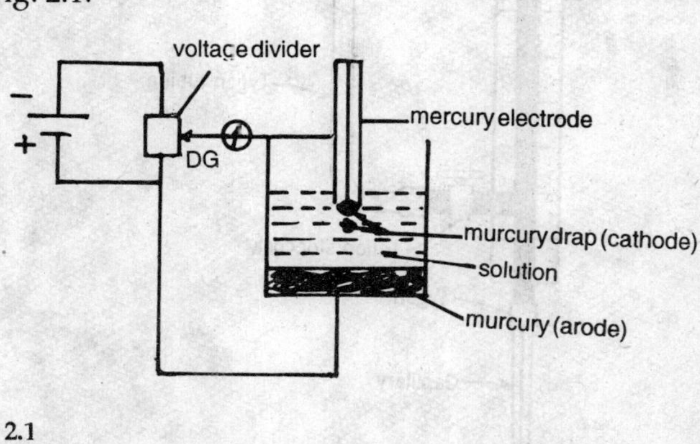

Fig. 2.1

Heyrovsky's Polarograph

It consists of a dropping mercury cathode in which a mercury drop grows steadily at the tip of a capillary dipping in the solution and falls. The surface of cathode thus remains fresh and the results are reproducible. Even at small current, the concentration polarisation is possible because of the small size of the cathode, drop falls slowly after every few seconds. It is called as an *indicator electrode* or *micro-electrode.*

INSTRUMENTATION AND PROCEDURE

Dropping Mercury Electrode (DME)

A typical arrangement for a dropping mercury electrode is shown is Figure 2.2. The drop time can be adjusted over a reasonably wide range by varying the height of the mercury reservoir. Useful drop times range from 2-9 sec; values shorter than 2 sec are used under special circumstances, and longer

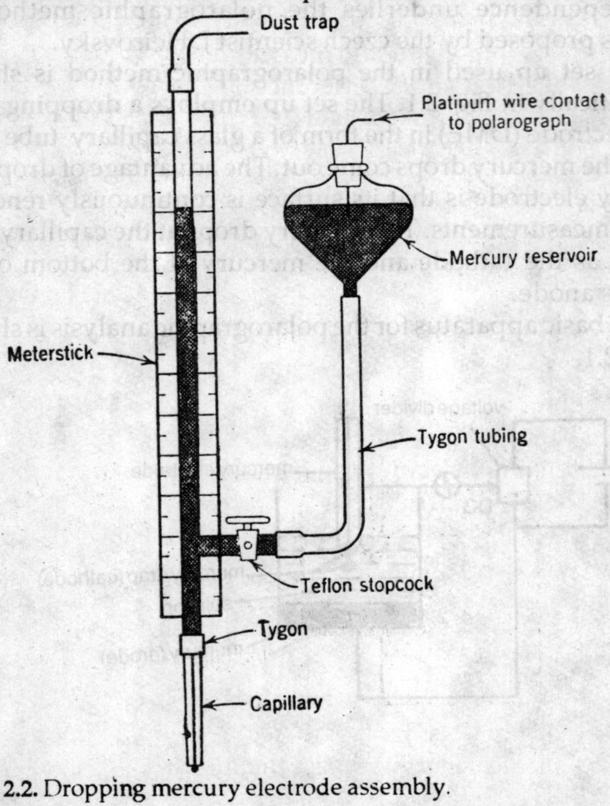

Fig. 2.2. Dropping mercury electrode assembly.

drop times produce on useful advantages. The capillary length is 5-10 cm; its internal radius is about 0.04-0.09 cm. Polarographic capillaries are commericially available for various drop time ranges.

Electrodes fabricated from materials other than glass such as Teflon are advantageous for use in corrosive solvents. With reasonable precautions, a capillary will retain its characteristics over long periods of use.

A DME is characterised by its capillary constants which are (i) mercury flow rate (m) and (ii) drop time (t),

The mercury flow rate (m) can be defined as the mass of mercury flowing through the capillary in are second (expressed is mg /s). The flow rate can be determined by allowing the mercury to flow through the capillary in weighed small beaker for 1 to 3 minutes. The beaker is weighed again and *m* is then calculated. These measurements are made at the same height of the mercury column at which polarographic study is normally carried out.

The drop time (t) is generally measured in 0.1 M kcl solution at a fixed applied potential difference on the DME against a reference electrode. It is then calculated from the time taken for 10-20 drops to fall from capillary at the same height of the mercury column at which *m* is calculated. The drop time depends on the composition of the solution as also on the voltage applied to the electrode. Generally the drop time ranges between 3-6 seconds. The drop time of DME can be changed by moving the bulb up or down.

The electric contact to the DME is made through the mercury filled in the levelling bulb using a noble metal (such as platinum) wire.

Anode : A poal of mercury located at the bottom of the call can serve as reference electrode and anodes. Because of its large area it is negligibly polarised.

Alternately standard reference electrode such as calomel electrode may be used.

The anodes selected or constructed are such that no concentration polarisation effects are in them.

Mercury used is DME must be pure and free from traces of foreign elements. The mercury be purified as follows before use. Wash it with pure water in order to extract soluble impurities present at its surface. For this shake the mercury with water in a separating funnel. After washing with water wash it with 1:5 dilute HNO_3 and finally the mercury is washed with distilled water.

To test the degree of purity of mercury use the **'foam test'**. Pure mercury if shaken with distilled water, forms a foam at its surface that remains for about 10 sec after keeping. Presence of impurities is indicated if the foam disappears rapidly or collapses suddenly. Dirty mercury does not foam at all.

The true potential of DME is related to observed potential by the following relationship.

$$E_{obs} = E_{dme} + iR$$

Although different microelectrodes like rotating platinum electrode, stationary nobel metal electrodes etc., have been used in polarograph, yet the DME is the best for obtaining current voltage curves because of the following advantages:

(i) Its surface area is reproducible with any given capillary.

(ii) Mercury forms analgams with many metals so it lowers their reduction potential.

(iii) It is possible to calculate the surface area from the weight of the drops.

(iv) The diffusion current assumes steady current quite instantaneously and is reproducible.

(v) High over voltage of hydrogen on mercury makes possible the deposition of ions which are difficult to be reproduced.

The DME is found useful in the range +2.4 v to - 2.8 v.

There are certain disadvantages of DME, some of these are as under:

(i) Mercury is quite toxic and should be handled carefully.

(ii) The cathodic versatility of mercury is not matched by its anodic behaviour. Mercury dissolution setting in at about 0.4 v (saturated calomel electrode). For metal ions it is immaterial because they get reduce at considerably lower potentials.

(iii) Oxygen is reduced by two stage process that shows a current voltage curve extending over most of cathodic working range.

At high current densities the non-polarisable electrode is unable to maintain a constant potential. In such cases the polarograms get distorted or drawn out.

This difficulty can be overcome by using a *three electrode assembly* which includes a DME, a larger counter electrode and a reference electrode. The potential difference is applied between the DME and the counter electrode while the potential of DME is evaluated against reference electrode; there is no flow of current between the DME and reference electrode. The DME and the reference electrode are kept as near as possible to minimise the solution resistance. In modern polarographs, provision is made for connecting the three electrode assembly is such a way that the true potential of DME is continuously monitored during the recording of the polarograph.

Inert Atmosphere

The solution prepared in solution contain nearly 10^{-3} M dissolved oxygen at room temperature. The dissolved oxygen gives two polarographic reduction waves over a wide range of working potential which interferes with most of the substances to be analysed polarographically. Oxygen can either be removed physically by bubbling an inert gas through the solution or chemically by adding a reducing agent, such as sodium sulphite, to the solution. Pure nitrogen and hydrogen are generally used as inert gases for removing oxygen from solution.

The dissolved oxygen in an aqueous solution can also be reduced chemically by adding sodium sulphite. The felt of such a solution must be kept higher than 8. A 0.2 ml saturated solution of sodium sulphite will be adequate for the reduction of dissolved oxygen present in a 10 ml solution. Although physical removal of oxygen by bubbling hydrogen or any other inert gas is the preferred method, chemical reduction by sulphite is found convenient in some estimations using amperometric titrations.

Solvent Composition

The vast majority of polarographic work is performed in aqueous solution. The inert supporting electrolyte concentration should be at least 100 times that of the ions to be determined. Alkali metal salts are commonly employed. The reduction potentials of sodium and potassium ions are – 1.8 and – 1.9 V vs SCE, allowing observation of most common inorganic species. More difficulty reduced species require a supporting electrolyte with a more cathodic reduction potential such as the tetraalkylammonium cations. An example is tetramethylammonium bromide (reduced at – 2.6 V vs SCE). However, the quaternary ammonium salts sometimes exhibit undesirable surface-active properties.

For organic compounds, nonaqueous solvents are often required to attain solubility of the sample. Some common solvents are acetonitrile, alcohols, dimethyformamide and acetone. Nonaqueous systems pose certain problems, such as choice of appropriate supporting electrolyte (and buffer for a organic sample) and high solvent resistance (with attendant *iR* drop), and any new system must be approached with caution.

Dissolved oxygen is reducible at the dropping mercury electrode, producing irreversible waves at –0.1 (H_2O_2 product) and – 1.0 V (OH - product) vs SCE. These waves interfere with most polarographic reductions, and oxygen must be removed rigorously from the sample solution prior to determination of a polarogram. This is accomplished by bubbling a stream of inert gas (such as nitrogen, hydrogen, or argon) through the solution.

Instruments

The basic requirements for a polarographic instrument are (a) a variable source of potential to cover the range 0 – 3 V, and (b) a method for measuring the current flow, which is usually less than 100 μA. It is desired that the applied voltage be known to within ± 0.001 V and that the maximum sensitivity of the current measuring device should be at least 0.01 μ A. Polarographic instruments can be of the "manual" type (in which the potential setting and current measurement are accomplished manually) or of the recording type (in which a linearly increasing potential is applied to the cell and the current flow is measured automatically).

A typical manual circuit is shown in Figure 2.3. The battery voltage is applied across the linear voltage divider R_1 (50–100Ω), and the cell voltage is selected by the setting of the variable contact. The current is measured in terms of the iR drop across the precision resistor R_2 in series with the cell by means of the potentiometer. By reversing the DPDT switch S, the applied

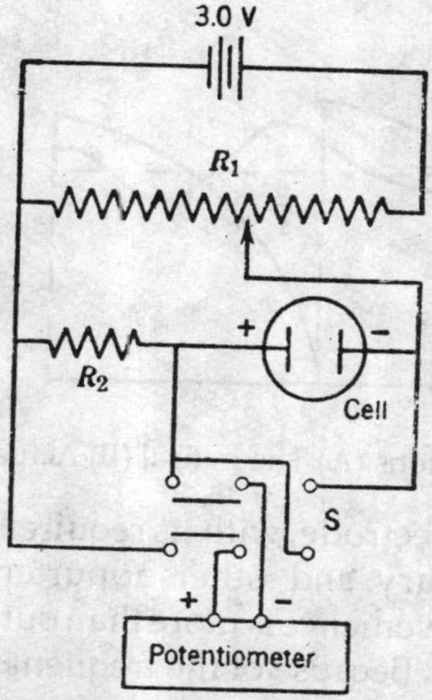

Fig. 2.3. Typical Manual polarograph.

potential may also be accurately measured using the same potentiometer. Some examples of commerical manual polarographs manufactured in the United State are the Fisher Scientific Co. Electropode, the Cambridge Instruments Company Voltammoscope, the Sargent Manual Polarograph Model III, and the American Instrument Company Polarometric Analyser.

Current Measuring Devices

Current is generally measured with the help of a sensItive galvanometer. The galvanometer can be calibrated at different sensitivity settings.

If a moving coil galvanometer is used its readings oscillate between maximum and minimum limits during the growth and fall of the drop of the DME. A galvanometer with fast response can almost follow the current variation during the growth of a drop. Fig. 2.4 shows the variations in current as followed by a fast response galvanometer (line A). A damped or slow galvanometer cannot follow various changes faithfully (line B).

Conventionally the galvanometer deflection is taken as positive during a reduction process. It is taken as negative during an ovidation reaction.

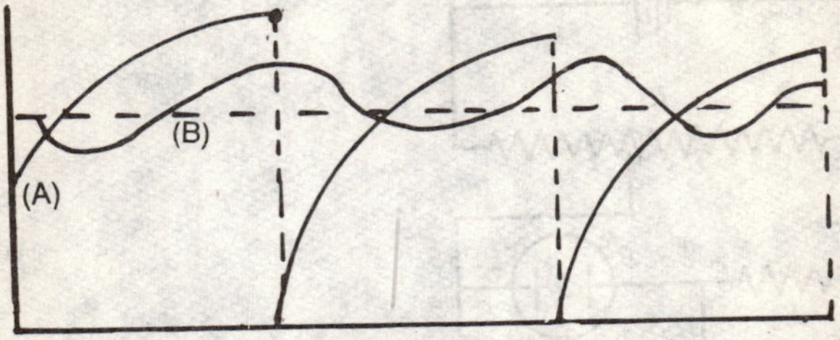

Fig. 2.4. Galvanometric Oscillations (A) Theoretical (B) Actual.

The dropping mercury electrode, with its requirement of an elevated reservoir of mercury and other appurtenances, is rather nuisance, but its inconvenience is more than outweighted by several major advantages. Because of the frequent dropwise renewal of the electrode, the concentration of any reduced metals dissolved in the mercury can never reach significant levels; also contamination of the electrode surface by adsorption cannot accumulate beyond the life of each drop. In common with all forms of mercury electrodes, the DME exhibits high over voltage with respect to the reduction of water to hydrogen. This latter feature makes accessible to polarography ions of many metals with reduction potentials in the range extending up to about – 2.0 V versus SCE, that could never be observed at platinum electrode. (However, platinum is advantageous at more positive potentials where mercury would be oxidized).

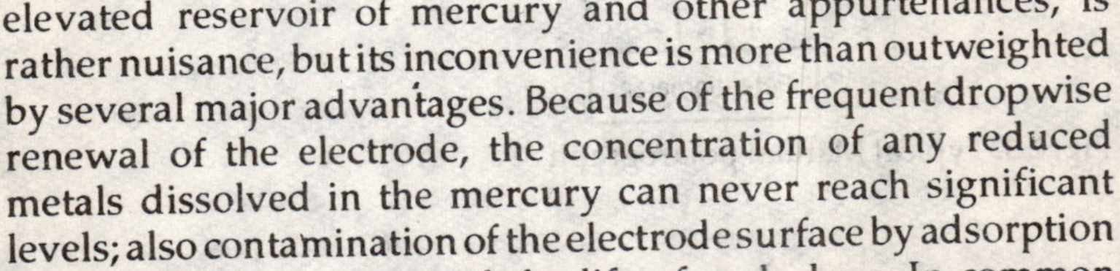

Fig. 2.5. A simple polarograph with a three-electrode cell containing a Pt-wire anode (AUX), a DME cathode, and an SCE as reference.

The Basic Experiment

Consider an unstirred, three-electrode cell (Figure 2.5) connected to a potentiostat. The working electrode is the DME; the reference is assumed for this discussion to be a Saturated Calomel Electrode (SCE). The auxiliary (counter) electrode can be a mercury pool or a platinum wire. The DME capillary is connected by a flexible tube to an elevated reservoir of mercury. The electronic circuitry is so designed that the potential advances stepwise, each drop being operated at a potential some 10 mV more negative than the previous drop. Each drop is knocked off mechanically at the same moment that the voltage is incremented, so that the drop is maintained at a constant potential throughout its lifetime.

Let us suppose that the cell contains a 0.5M solution of KNO_3 saturated with air. The current passing through the cell is recorded as a function of the voltage. The resulting graph is shown is Figure 2.6 a, curve 1. The deep serrations are the result of successive drop falls, in which the surface area, and hence the current, goes almost to zero. By an added electronic component called a *sampling circuit*, the serrations can be eliminated, giving curve 2, in which a series of steps correspond to sequential drops. This is desirable for simplicity, since no additional information is carried by the lower portions of the serrations.

The waves in these curves are due to the cathodic reduction of dissolved oxygen, as can be shown by running similar curves after flushing out the oxygen with an inert gas, such as purified N_2 or Freon-12, a process known as *sparging*. Figure 2.6 b demonstrates the change in the curve resulting from oxygen removal. (Because of the effect, it is standard practice to sparge all solutions routinely before runnning polarograms.)

If now the amplifier gain be increased, a response such as curve 1 (unsampled) of Figure 2.7 *a* will result. Since no electroactive substance is present, the curiously shaped serrations must be due solely to the double-layer charging effects. The marked change at about – 0.57 V coincides with the PZC, the point of zero charge, where there is no tendency to form a charged double layer, and hence where no current passes. At points on either side of this, transient currents flow to charge the capacitance of the double-layer, but with opposite signs. As each drop falls from the capillary, the accumulated charge is abruptly lost and a new transient commences. If this is sampled

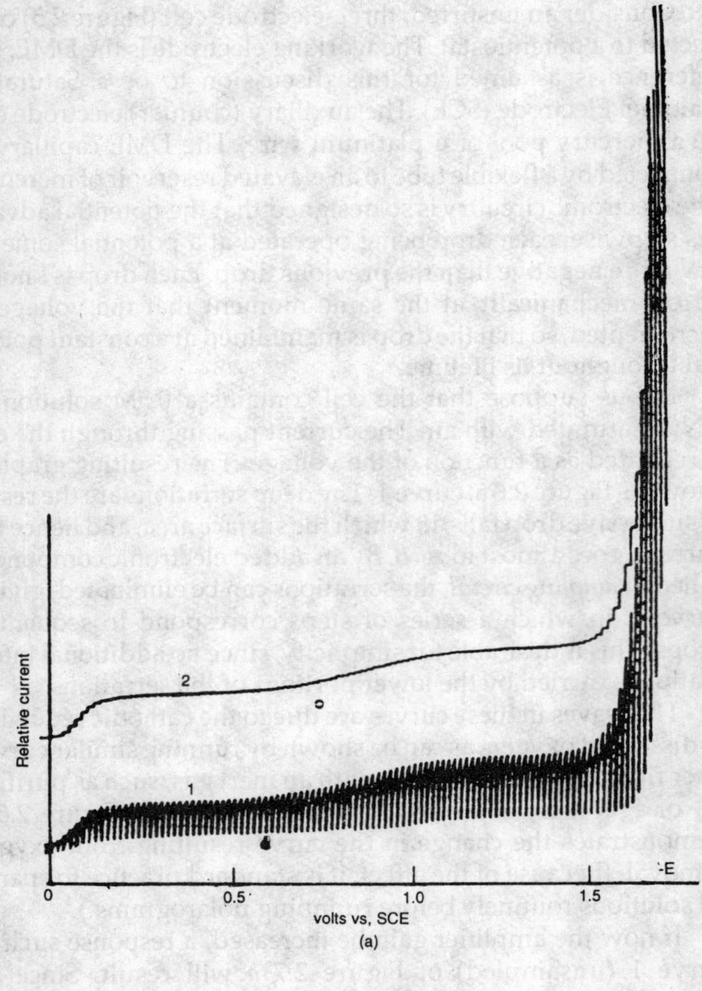

Fig. 2.6. (*a*) Polarogram of supporting electrolyte 0.5 M KNO_3) in the presence of air-curve 1, conventional recording, curve 2, sampled (moved upward for clarity). (*b*) The effect of sparging with Freon-12 for various time intervals as marked; solution: 0.5 *M* KCl. (Recorded on a Sargent-Welch Polarograph).

just prior to each drop fall, curve 2 is seen, which is effectively the *upper* envelope of the more positive segment and the *lower* envelope of the more negative portion of the curve.

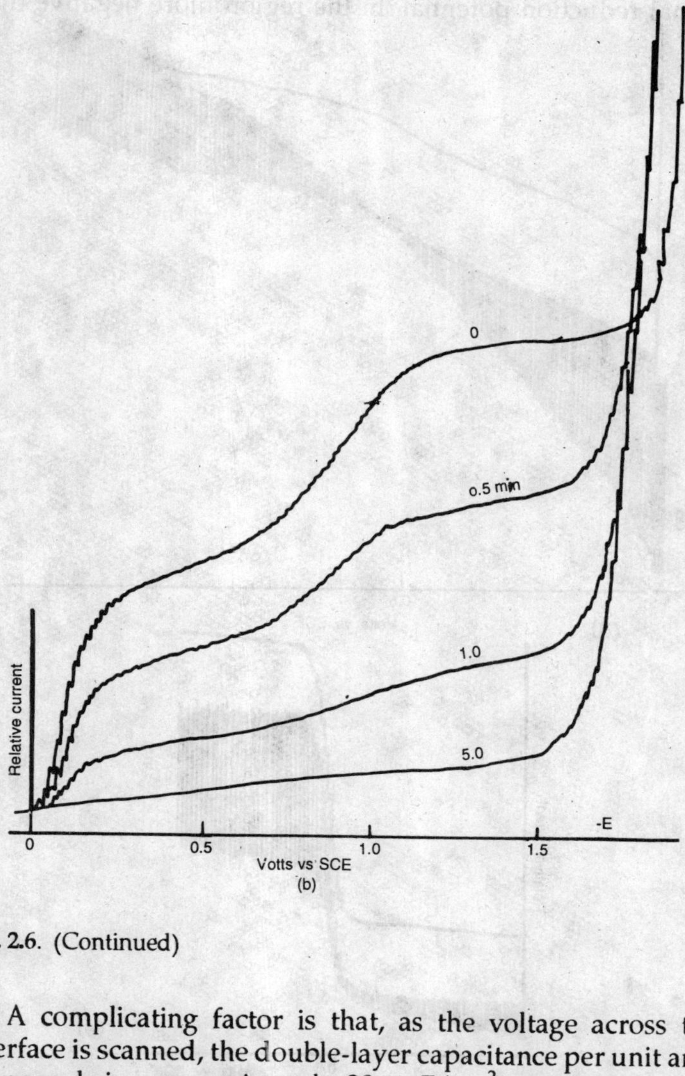

Fig. 2.6. (Continued)

A complicating factor is that, as the voltage across the interface is scanned, the double-layer capacitance per unit area changes, being approximately 20 μ F/cm^2 at voltages more negative than the PZC, and about twice this value at more positive potentials. This causes a corresponding variation in the charging current.

If an electroactive substance, such as Pb^{++} is present, the curves become modified as shown in Figure 2.7*b*. An abrupt step (usually called a "wave") appears at a voltage near to the formal reduction potential. In the region more negative than

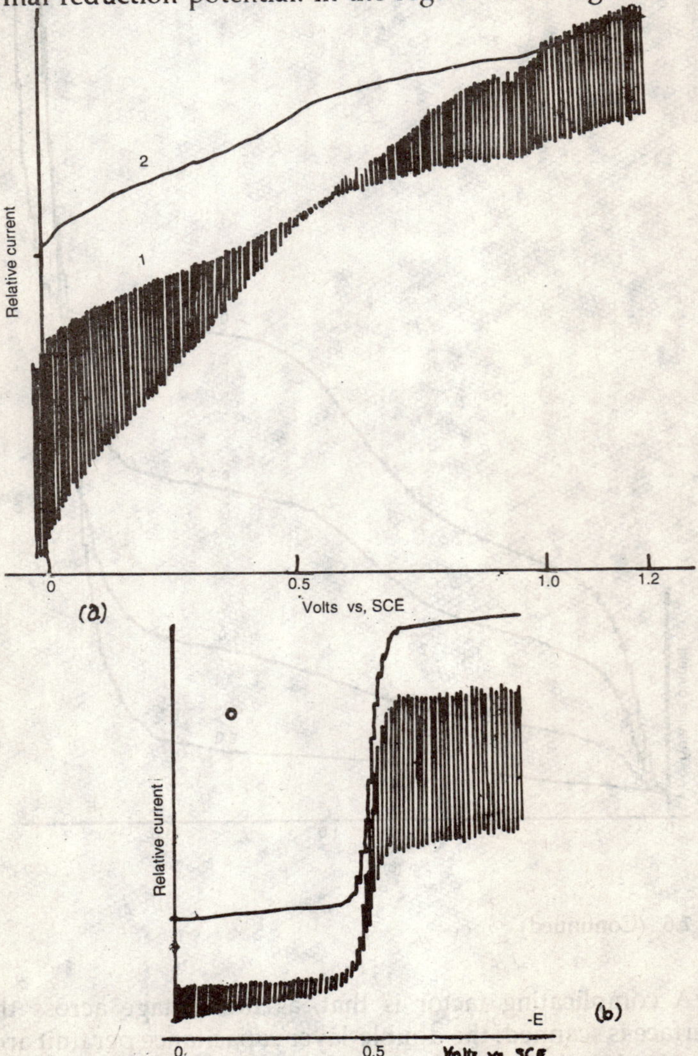

Fig. 2.7. Typical DC polarograms. (*a*) Curve 1: unsampled background current; Curve 2, same, sampled. (*b*) 10^{-4} M Cd^{++} in 0.5 *M* KCl, sampled and unsampled. Sparged with Freon-12. Drop-time mechanically regulated at 1 s. Scan rate, 0.5 V/min. (Recorded on a Sargent Welch Polarograph).

about –0.4 V in this example, a new component of current is evident, originating in the faradaic reduction of lead ions to the free metal. It may be noted that the sampled curve now corresponds to the *upper* envelope of the unsampled counterpart over the whole range.

An important innovation in polarographic instrumentation was announced in 1979 by Princeton applied Research Corporation. This firm introduced a mercury drop electrode assembly in which the mercury flow can be interrupted by a fast-acting valve under the control of an electronic timing circuit. The mercury is allowed to flow for a short period, such as 100 ms, following which the drop remains at constant size, hanging from the tip of the capillary by surface tension until dislodged and replaced. The advantage of this system (called the SMDE, for Stationary Mercury Drop Electrode), is that a measurement can be made on a drop of constant surface area. This avoids the difficulties inherent in the large periodic double-layer charging current of the conventional electrode, while preserving the advantage of renewing the mercury drop. Figure 2.8 shows the graph of drop area against time for the two types of electrodes.

The mathematical theory, as well as practical operating procedures, are considerably simplified by substitution of the SMDE for the traditional DME.

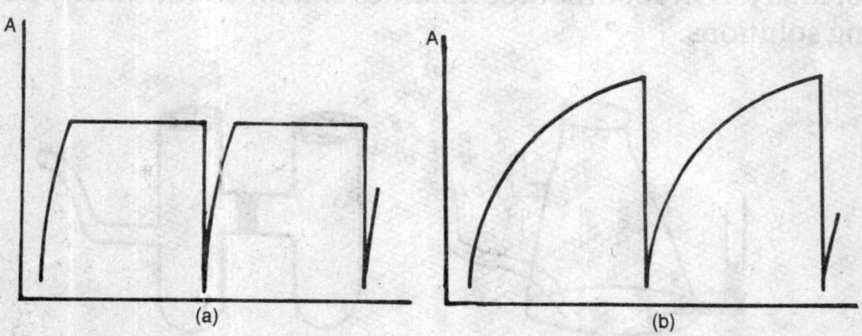

Fig. 2.8. The area of mercury-drop electrodes as a function of time. (*a*) Stationary mercury drop (Princeton Applied Research Corporation). (*b*) Conventional DME.

With the advert of operational amplifiers it has become possible to perform mathematical operations easily which are needed in handling electronic signals. It is now possible to fabricate polarographs with the help of the operational amplifiers if one has a basic knowledge of their function.

A polarograph constructed with the help of operational amplifiers can amplify currents without loading the system with undesirable voltage drop.

Recording polarographs are very convenient to operate and draw i–E curves continuously on which any abnormality can be easily detected.

Polarographic Cells

Two general type of cells are used for polarographic analysis: Those with internal reference electrodes whose potential depends upon the solution composition (a mercury pool or silver wire) and those employing an external reference electrode whose potential is independent of the solution composition. The former are used where accurate E½ (Half wave potential) values are not required, such as in routine analysis, and are well behaved only if the solution contains an anion of a slightly soluble salt of the anode metal (such as chloride ion for mercury) in an oppreciable concentration. Typical cell construction is shown in Fig. 2.9 (a). External reference electrodes are more commanly used. A typical H-cell is shown in Fig. 2.9 (b), the refrence electrode is separated from the sample by an agar plug held in place by a scintered glass disc. Polarographic cells are commercially available in many forms including modifications for analysis in small solution slumes and for continuously flowing solutions.

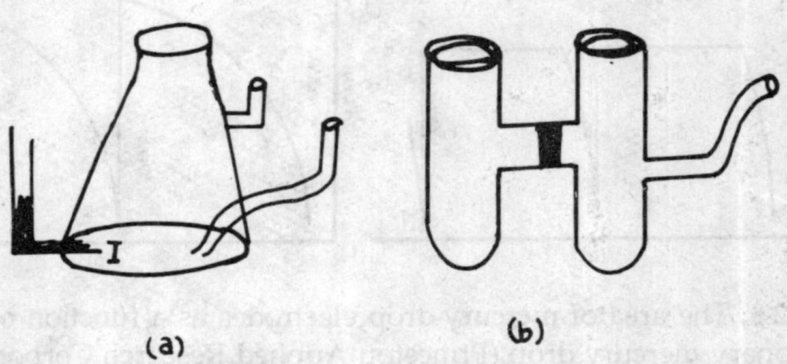

Fig. 2.9. Polarographic cells (a) for internal mercury pool reference electrode and (b) for (isolated) external reference electrode.

Types of Polarographs

There are various types of polarographs such as (i) Manually operated polarographs and (ii) Automatic polarographs which record current voltage curve automatically.

It is possible to record photographically on a sheet of sensitized paper which is fixed to a drum which can be rotated by a small motor.

Procedure: A simple apparatus (Fig 2.10) for polarography consists of DME, capillaries reference cell, rheostate, shunt, galvanometer, battery and standard resistor. The solution to be analyzed is placed in the cell and the dropping mercury electrode is introduced in it so that the lower end of the capillary in immersed in the solution.

The dropping mercury electrode is adjusted in such a manner that the mercury leaves the capillary at the rate one drop per 2 to 6 seconds. Mercury is easily purified and in this arrangement its surface is continually renewed as the drop grows. An external reference electrode completes the electrolysis cell. The electrical circuit provides a source of continuously variable applied voltage and a device for measuring the very small electrolysis current (10^{-8} to 10^{-4} amp.) On modern instruments, the contact is driven across the slide-wire by a synchronous motor which increases the applied voltage at a constant rate; and the current is measured by means of an automatic recorder. (see Figure 2.10).

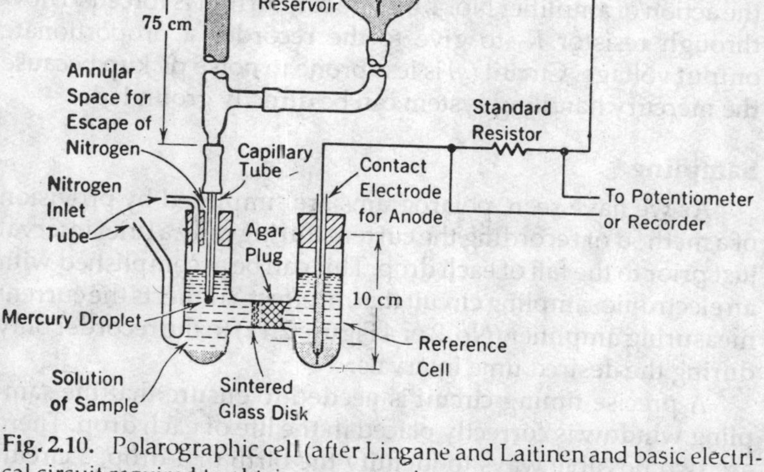

Fig. 2.10. Polarographic cell (after Lingane and Laitinen and basic electrical circuit required to obtain current-voltage curves.

Polarography is the study of current voltage relationships obtained with this apparatus. With the proper conditions, the **polarogram** (current vs voltage) so obtained can be interpreted to yield both the nature and concentration of reacting species.

DISCUSSION OF A D.C. POLAROGRAPH

Instrumentation

A DC polarograph must contain a unit for producing a voltage ramp to be applied to the cell, a potentiostat, and a current-measuring device. Modern instruments are automatically recording and include a ramp generator that is synchronized with the recorder.

Figure 2.11 gives block diagrams of two widely used versions of polarographic instrumentation. The unit marked "ramp generator" consists of either an electronic integrator that gives a potential $E=k\int dt = kt$, or a step function generator that produces a staircase approximation to a ramp. In either case, its output is an increasing voltage, adjustable in both range and slope. This ramp voltage is tracked by the potentiostat (amplifier No. 1 or 3) and impressed upon the solution as monitored by the reference electrode. In circuit *(a)* the DME is directly connected to ground, and the current is measured by the voltage drop across resistor R_1 in series with the auxiliary electrode, as sensed by the differential amplifier, No. 2. In the alternative circuit at *(b)*, the DME is held at *virtual* ground by the action of amplifier No. 4, but the cell current is forced to flow through resistor R_2 to give to the recorder a proportionate output voltage. Circuit *(a)* is less prone to noise pickup because the mercury handling system can be directly grounded.

Sampling

As we have seen, polarograms are simplified by provision of a method of recording the current during only a brief interval just prior to the fall of each drop. This can be accomplished with an electronic sampling circuit that, in effect, connects the current measuring amplifier (No. 2 or 4 Figure 2.11) to the recorder only during the desired time intervals.

A precise timing circuit is needed to ensure that the sampling window is correctly placed in the life of each drop. There are two possible ways to identify the birth of a drop; a circuit

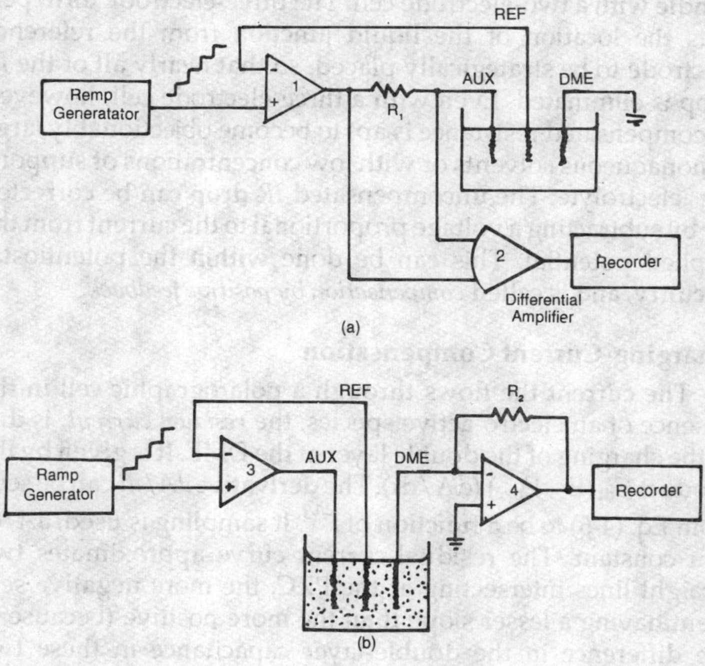

Fig. 2.11. Simplified schematic diagrams of two common polarographic circuits. In (a) the DME is directly grounded, thereby reducing electrical noise pick-up. Circuit (b) employs a current-to-voltage converter (amplifier # 4).

can be arranged to detect the fall of the previous drop and start the timing sequence, or an electromechanical drop-knocker can be provided to cause the mercury to fall at preselected intervals. The latter method is usually preferred, since it ensures an adjustable, uniform drop time throughout an experiment. Timing circuits for either option are easily designed with standard electronic components.

Resistance Compensation

Under some circumstances, the resistance of the solution may become significant. This tends to increase the uncertainty of the actual potential across the interface, since the drop in voltage produced by the resistance of the solution changes with the current that is being measured. This is especially difficult to handle with a two-electrode cell. The three-electrode form permits the location of the liquid junction from the reference electrode to be strategically placed, so that nearly all of the *IR* drop is eliminated. Even with a three-electrode cell, however, uncompensated resistance is apt to become objectionably large in nonaqueous solvents or with low concentrations of supporting electrolyte. The uncompensated *IR* drop can be corrected for by subtracting a voltage proportional to the current from the applied potential. This can be done within the potentiostat circuitry, and is called *compensation by positive feedback.*

Charging-Current Compensation

The current the flows through a polarographic cell in the absence of an electro-active species, the *residual current,* is due to the charging of the double layer at the DME. It is given by the product C_{dl} $(E—E_{pzc})$ (dA/dt). The derivative *dA/dt* can be seen from Eq. (4-6) to be a function of $t^{-1/3}$. If sampling is used, *dA/dt* is a constant. The residual-current curve approximates two straight lines intersecting at the PZC, the more negative segment having a lesser slope than the more positive (because of the difference in the double-layer capacitance in these two regions).

The slope of the residual current causes difficulty with dilute solutions because of the necessary increase in amplification. The situation can be improved by a procedure known as *linear baseline compensation,* which consists of adding to the output an opposing voltage ramp just sufficient to return the residual current trace to an approximately horizontal line. The compensation can never be perfect, but it can nevertheless be very helpful. Figure 2.12 shows an example of its use.

A polarograph with all the foregoing features is diagrammed in Figure 2.13.

Cell Design

Much of the early work in polarography was carried out with a simple two-electrode cell in which a large pool of mercury served as combined auxiliary and reference electrode (Figure 2.14a). It was necessary for the supporting electrolyte to contain an "anodic depolarizer," such as chloride or other ion

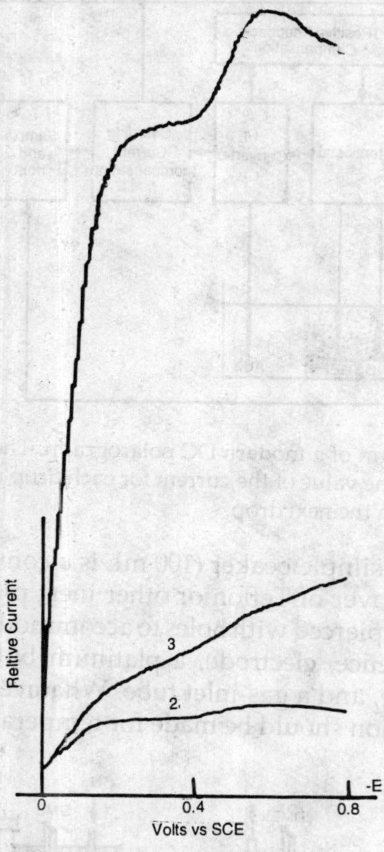

Fig. 2.12. Sampled DC polarograms of 0.64 ppm thallium in 0.5 *M* KCl. Curve 3: without compensation; the wave at E½=0.45 V is hardly discernable. Curve 2: same solution, but with linear compensation; the reduced background slope makes it possible to increase the gain 10 times to give Curve 1, from which a fairly precise measurement of the wave height can be made. (Recorded on a Sargent-Welch Polarograph).

that forms an insoluble salt with mercury, to provide a stable reference potential..

The two-electrode design reached its zenith with the H-cell (Figure 2.14*b*), wherein the reference electrode is permanently attached by a connecting salt bridge closed with a glass frit and filled with a KCl-containing plug of agar gel.

Since the general introduction of the three-electrode cell concept, the H-cell has become obsolescent. More rugged and

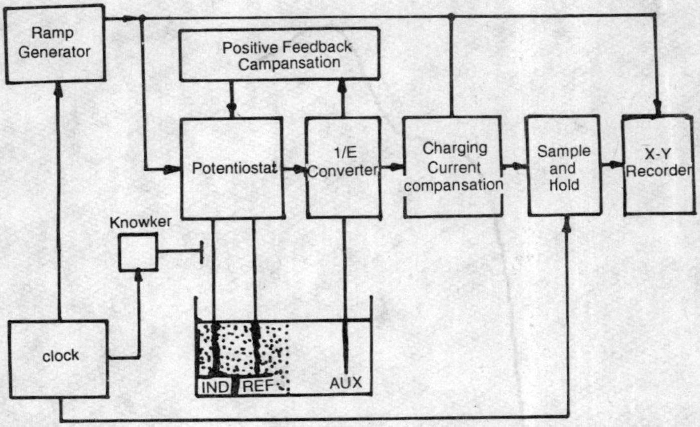

Fig. 2.13. Block diagram of a modern DC polarograph. The sample-and-hold module retains the value of the current for each drop until up-dated with information from the next drop.

less expensive is a simple beaker (100 mL is a convenient size) provided with a cover of Teflon or other inert plastic (Figure 2.14c). The cover is pierced with holes to accommodate the DME capillary, a refference, electrode, a platinum button or wire auxiliary electrode, and a gas-inlet tube. Whatever cell design is adopted, provision should be made for temperature control.

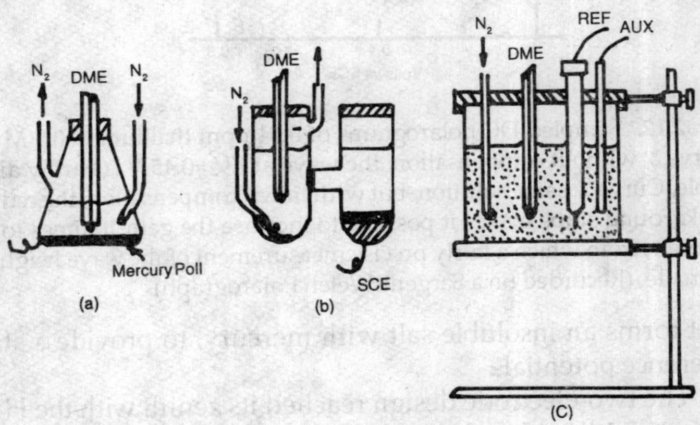

Fig. 2.14. Representative polarographic cells. (a) Heyrovsky design. (b) Lingane-Laitinen H-cell. (c) Suggested three-electrode cell with supporting cover; see also the frontispiece.

The Capillary System

A typical glass capillary through which mercury is delivered to form a DME is about *50 μm* internal diameter and 15 cm in length. It is important that it be supported vertically and that the lower end be cut squarely. The capillary is connected to reservoir by flexible tubing. The capillary for use with the SMDE is shorter and with larger bore to allow a drop to form quickly.

A capillary, if properly cared for, should last for many months. A basic precaution that must be taken is to prevent any dissolved solid matter from entering the capillary opening. Another precaution is never to allow the potential to become so positive as to permit the anodic oxidation of mercury. Some of the Hg_2Cl_2 or other salt so formed is certain to get into the capillary and render it useless. The DME is quite sensitive to vibration, and the location of the instrument should be slected carefully with this in mind.

EXPERIMENTAL METHODOLOGY

Supporting Electrolytes and Half-Wave Potentials

A great variety of materials have been used as supporting electrolytes. Some are listed in Table 2.1. The choice is sometimes not at all easy, particularly if more than one active species is present. Table 2.2 lists a selection of half-wave potentials in the electrolytes mentioned. Study of this table suggests that species which interfere in one medium can often be resoved in another. For example, Sn(IV) and Zn(II) are reported to have identical $E_{1/2}$ values in acetate, whereas they are widely separated in H_3PO_4. As another example, suppose that two waves are obtained in HCl, with $E_{1/2}$ about –0.43 and –0.65 V. This combination might be ascribed to the double wave of As (III) or just as well to a mixture of Pb (II) and Cd (II). A shift to an acetate medium would clarify the situation.

It is possible to gain much information about complex formation, including equilibrium constants, from polarographic studies. The $E_{1/2}$ for the reversible reduction of a complex to a metal soluble in mercury is displaced toward more negative potentials as compared to the uncomplexed hydrated metal ion. If both oxidized and reduced species are water-soluble and form complexes, the situation becomes more involved, and the shift of potential can be in either direction. In irreversible systems, the effect of complex formation may also be quite diverse.

TABLE 2.1
Supporting Electrolytes for Polarography

	For Aqueous Solutions
1. Neutral	$NaClO_4$ or KNO_3
2. Acetate	2 *M* HOAc + 2 *M* NH_4 OAc
3. Ammonia	1 *M* NH_3 + 1 *M* NH_4 Cl
4. Chloride	0.1 *M* KCl or KaCl
5. HCl	1 *M* HCl
6. Citrate	Citric acid + Ammonium citrate, pH=4
7. Citrate	Same, pH=6
8. EDTA	0.1 *M* EDTA, pH=7
9. Hydroxide	1 *M* KOH or NaOH
10. Oxalate	0.25 *M* Oxalic acid + $(NH_4)_2$ – Oxalate, pH=4
11. Phosphat	7.3 *M* H_3 PO_4
	For Nanaqueous Solutions
LiCl, Me_4NIO_4, Bu_4 NI, $LiClO_4$, according to solubilities.	

It is because of the great variation of half-wave potentials from one electrolyte to another that polarography is seldom a technique of choice for *qualitative* analysis. It can, of course, serve negatively to establish the *absence* of some element. A more general qualitative scheme for the elements would necessitate running the same sample in various supporting electrolytes and correlating the data so obtained. For more than three or four ions, this would become too combersome. Qualitative identification of organic compounds is likely to be even less productive.

Resolution

Resolution in this context refers to the needed separation of two waves. The basic requirement is that the half-wave potentials be sufficiently far apart that the plateau between them will become nearly paralled to the baseline. Generally, two reversible waves of equal height and the same *n*-value may be con-

TABLE 2.2
Half-Wave Potentials of Elements in Aqueous Solution

Supporting Electrolytes	As^{III}	Cd^{II}	Co^{II}	Cr^{IV}	Fe^{II}	Fe^{III}	In^{III}	Pb^{II}	Sn^{II}	Tl^{I}	Zn^{II}
1	–0.07 –1.00	–	– *	–	* –1.46	>0	–	–0.38	* (+0.14) –0.43	–	–1.00 *
2	–0.92 *	–0.65	–1.19	–	NR	>0	–0.71	–0.50	(–0.16) –0.62	–0.47	–1.10
3	–1.41 –1.63	–0.81	* –1.29 *	–0.20 –0.60 –0.30*	–0.34 –1.49 *	ppt	–	ppt	–	–0.48	–0.20 –1.60
4	– *	–0.60	–1.20	–1.00 –1.80	–1.30	–	–0.56	0.40	–	–0.46	–1.00
5	–0.43 –0.67	–0.64	NR	–	NR	>0	–0.56	–0.44	(–0.10) –0.47	–0.48	NR *
6	–0.75	–0.59	NR	–	(–0.05)	+0.05	–	–0.43	(–0.21) –0.54 *	–0.45	–1.04
7	* –1.46	–0.70	NR	>0 –0.36	(–0.18)	* –0.23	NR	–0.56	(–0.41) –0.66	–0.51	* –1.37
8	–	–1.27 *	NR *	– *	–	–1.15	–1.08	–1.37	–1.26	–0.48 –0.48	NR
9	(–0.26)	–0.78	–1.46	–0.85	(–0.90)	–	–1.09	–0.76	(–0.73)	–1.22	–1.53
10	–0.79 –0.46	–0.63	– *	–	(–0.23)	–0.23	–	0.50	–0.70 *	–0.46	– *
11	–0.71	–0.77	–1.20	>0	NR	+0.06	–	0.53	–0.58	–0.63	–1.13

Notes : *Irreversible; NR not reduced; – data not available; ( ) anodic wave; two or more entries, multiple wave; supporting electrolytes.

sidered resolved if their half-waves potentials are separated by 300/*n* mVMv, but this is only a guide rather than a firm criterion. For irreversible reduction, the required separation is considerably greater.

Voltammetry : 1. Polarography

It must be realized that the number of electroactive species that can be determined in a given supporting electrolyte is severely limited by this resolution requirement. At equal concentration, for $n = 2$, the maximum number of waves theoretically resolvable is 2/0.15=13, where the available range is taken as 0 to –2 V. Of course, this number cannot be attained in practice because half-wave potentials are not evenly distributed over the range.

The separation necessary for adequate resolution depends also on the relative heights of the waves. This effect is likely to be more important where two adjacent waves are greatly different in height; the smaller wave is likely to be swamped out by the larger.

Precision, Accuracy and Sensitivity

As is any analytical method, it is desirable to estimate the precision that it is reasonable to expect. Fisher has considered this carefully for the case of DC polarography and concludes that quantitative measurement of a 5×10^{-3} *M* solution of Cd^{++} can easily reach a precision of 2 per cent of the amount present. This can be improved by perhaps an order of magnitude with special precautions.

The accuracy of a determination, of course, depends not only on the precision (reproducibility), but also on the accuracy with which the concentrations of standards are known.

The sensitivity of the polarographic method is limited by the magnitude of the charging current. The practical limit of useful measurements is reached with about 10^{-4}*M* solutions for the conventional (DC) method.

Currents in Polarography

In discussion of methods of polarography we have already noted that when an external e.m.f. is applied all the positively charged ions will be attached to the dropping mercury electrode by an *electrical force* and by a *diffusive force* which results from the concentration gradient formed at the surface of the electrodes. The total current can be considered as equal to the sum of electrical force and diffusive force.

When the applied voltage is increased and the current is recorded, the plot obtained as shown in Fig 3.1 is for a solution containing a small amount of (~10^{-3} *M*) Cd^{2+} ions (reducible substance) in a conducting medium (1.0 *M* Kcl).

A significant increase occurs at *A* when applied potential difference reaches a value when the following reaction is initiated

$$Cd^{2+} + 2e^{-} \rightarrow Cd\,(Hg)$$

The current increases sharply and reaches a limiting value at B where DME potential becomes so negative that all the cadmium ions reaching the electrode per unit time are reduced. The rising portion of the curve from A to B assumes a S-shape with a point of inflexion at $E_{1/2}$. Further increase in negative potential has no significant effect on limiting current. On reaching a sufficiently high negative potential at C, a rise in current occurs again which continues unabated.

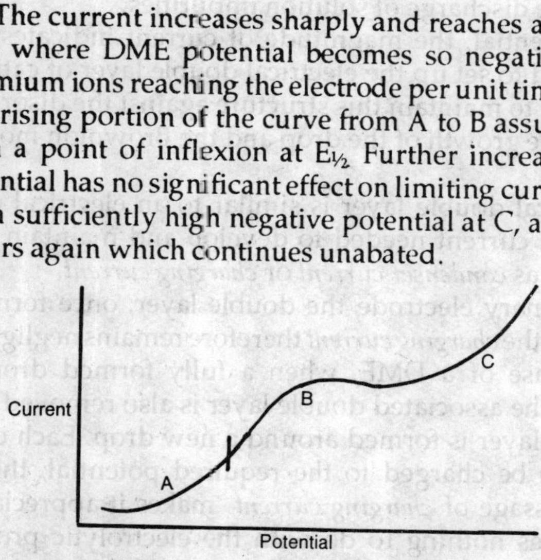

Fig. 3.1. Current potential curve.

During polarrographic analysis the measured current at any potential can be due to various factors which include.

(i) charging current or condenser current,
(ii) migration current,
(iii) diffusion current,
(iv) adsorption current,
(v) kinetic current, and
(vi) catalytic current.

It is not essential that all kinds of currents be present togather. By suitably adjusting the conditions many of these can be eliminated and the observed current can be validly interpreted:

Various kinds of currents are briefly discussed in the following pages:

(i) Condenser Current

The current flowing between the electrodes when the applied voltage is lower than the decomposition voltage is *A* called *residual current*. This current is called by the same electrode process as in the case of voltage higher than decomposition voltage, but at a low rate due to high electrode polarisation and also by deposition of impurities at the electrodes. At voltage lower than equilibrium potential difference the residual current is caused by the discharge of solution impurities.

At any potential, the magnitude of current indicates the energy required to set up the electrical double layer of catious and anions and to maintain this structure against the disordering effects of the growth of the drop and the Brownion motion in solution.

The electrical double layer is similar to an electrical condenser and the current needed to develop and maintain this layer is known as *condenser current* or *charging current*.

With stationary electrode the double-layer, once formed, remains stable, the *charging current* therefore remains negligible. However in case of a DME, when a fully formed drop of mercury falls, the associated double layer is also removed and a fresh double layer is formed around a new drop. Each drop must therefore be charged to the required potential, thus a continuous passage of *charging current* makes it appreciable. This current has nothing to do with the electrolytic process governed by Faraday's law. The magnitude of this current at the DME is a few tenth of a microampere in an air-free solution of potassium chloride or sodium perchlorate.

(ii) Migration Current

There are two processes by which electroactive materials reach the surfaces or respective electrodes.

(a) one involving the movement of charged particles in the electric field caused by the potential difference that exists between the electrode surfaces and the solution.
(b) the other involving diffusion of particles.

The current needed for both these processes is called *migration current.* This current can be totally eliminated by adding indifferent *electrolyte* to the solution. The concentration of indifferent electrolyte added should be so large that its ions carry essentially all the current (migration current depends upon the transport number of ions). The indifferent electrolyte added is known as *supporting electrolyte.* Generally the concentration of supporting electrolyte added is 100 times that of the electroactive material.

Experimentally it has been found that during the reduction of a cation, the migration current increases the magnitude of the limiting current while the limiting current is decreased during the reduction of an anion. This can be explained on the assumption that the total limiting currents the algebraic sum of diffusion and migration currents.

$$I_{el} \text{ (cationic reduction)} = I_d + I_m$$
$$I_{el} \text{ (anionic reduction)} = I_d - I_m$$

Where I_e, I_d and I_m are limiting diffusion and migration currents respectively.

During reduction the cations will migrate towards the cathode because of the electrostatic attraction. Thus, the movement adds to the limiting current. On the other hand, the anions move away from the cathode because of the electrostatic repulsion and thus migration operates in apposition to diffusion. The limiting current is lowered in such a case. The following relationship is valid for limiting and diffusion current:

$$\frac{Id}{Ie}\text{(cationic reduction)} = 1 - T^{+}$$

$$\frac{Id}{Ie}\text{(anionic reduction)} = 1 + T^{-}$$

where I_d is the diffusion current obtained in the presence of large excess of supporting electrolyte and I_e is the limiting current observed in absence of any supporting electrolyte. T^+ and T^- are the transport numbers of cations and anions respectively. If the values of transport numbers are about 0.5, it is found that during reduction of a cation the diffusion current observed in presence of a supporting electrolyte is about one and a half times that of limiting current in the absence of supporting electrolyte.

To ensure that the migration current becomes negligible the concentration of supporting electrolyte be at least 100 times to that of the electroactive species. This helps in decreasing the resistance of solution in the polarographic cell and in minimising the iR drop across the electrodes.

(iii) Diffusion Current

When the rate of electrode reaction determined by the transport of the discharging particles to the electrodes, solution concentration in the layer near the electrode, outside the electrical double layer changes. The solution near the electrode is deplated with discharging particles and the concentration polarisation of the electrode occurs.

Diffusion current is directly proportional to the concentration of the substance being reduced or oxidised at the dropping electrode.

The diffusion current density is given by *fick's equation*

$$i = i_{diffusion} = Z.F.D.\ (C_o - C_5)\ \delta_{diffusion}$$

where Z = charge of diffusing ion

F = faraday constant

D = diffusion constant

C_o = Bulk Concentration of solution at the boundry with the electrical double layer.

$\delta_{diffusion}$ = thickness of diffusion layer

The equation for polarographic diffusion current was given by Iikovic in 1954 and is known as *likovic equation*. According to it

$$i_{diffusion} = 607\ n.\ D^{1/2}.C.\ m^{2/3}\ t^{1/6}.$$

where

$i_{diffusion}$ = average diffusion current in micro amperes during the life of a drop.

n = number of Faradays required per mole of electrode reaction.
D = Diffusion coefficient in $cm^{-2} sec^{-1}$
C = Concentration of substance in millimoles per litre.
m = rate of flow of mercury from DME in mg / sec.
t = drop time in seconds.

From this equation it is quite clear that the "observed diffusion current is directly proportional to the concentration of the electroactive material."

The above equation has been modified by *Lingane* and *Loveridge* by applying correction for curvature of the electrode surface. The modified equation is

$$i_{\text{diffusion}} = 607\ n\ D^{1/2}\ m^{2/3}\ t^{1/6}\ (1+39\ D^{1/2}\ m^{-1/3}\ t^{1/6})$$

The term in the bracket corresponds to difference between linear and spherical diffusion. The correction is of the order of 3-7% only and this modified equation is used only in case of accurate work.

The Ilkovic Equation

Since in a polarographic experiment, the current is controlled by the concentration of the electroactive species, it is desirable to establish a mathematical relation between these variables. To do this, it is expedient to deal first with the equation for diffusion to a stationary electrode.

If the electrode is planar and the system is diffusion controlled, the response to a large voltage step follows the Cottrell equation:

$$I = nFA(D/\pi t)^{1/2}\ C^*_{OX} \qquad (3.2)$$

This equation is valid only for potentials that are sufficient negative to ensure that the surface concentration $\underline{C}^S$=0. A more general equation can be obtained if we remove this restriction. In this case, the current is given by:

$$I = nFA\left(\frac{D}{mt}\right)^{1/2}\left(\frac{1}{1+\theta}\right)C^*_{OX} \qquad (3.2)$$

in which

$$\theta = \exp\left\{\frac{RT}{nF}(E - E_{1/2})\right\} \cong C^S_{OX}/C^S_{RED} \quad (3.3)$$

The relation between the exponential and the ratio of concentrations can be derived by taking the logarithm of each side of the Nernst equation and substituting $E_{1/2}$ for $E^{o'}$. The exact significance of $E_{1/2}$ will be explained shortly.

We will now show how this equation can be modified to apply to the DMF, in which the electrode area increases with time. It is assumed that the functional relation of Eq. (3.2) is retained except that the quantity A becomes variable.

The area of a mercury drop under gravity control, as a function of time, can be shown by simple geometry to be:

$$A_{drop} = (4\pi)^{1/3}\left(\frac{3mt}{\rho}\right)^{2/3} \quad (3.4)$$

where m is defined as the rate of flow of mercury from the capillary, and ρ *is its density. Inserting this value into Eq. (3.2) gives.*

$$I = nF\,(4\pi)^{1/3}\left(\frac{3mt}{\rho}\right)^{2/3} D^{1/2}_{OX}\,(\pi t)^{-1/2}\,C^*_{OX}\left(\frac{1}{1+\theta}\right) \quad (3.5)$$

$$I = (const)\,.\,nm^{2/3}\,t^{1/6}\,D^{1/2}_{OX}\,C^*_{OX}\left(\frac{1}{1+\theta}\right) \quad (3.6)$$

where the constant factor includes the density of mercury, ρ = 13.6 g. cm^{-3}, as well as the Faraday constant. A correction must be made to account for the contraction of the diffusion layer as the drop expands into the solution; this factor has been shown [8] to be $\left(\frac{7}{3}\right)^{1/2}$. This gives the relation:

$$I = 7.06 \times 10^3\,.\,mm^{2/3}\,t^{1/6}\,D^{1/2}_{OX}\,C^*_{OX}\left(\frac{1}{1+\theta}\right) \quad (3.7)$$

Equation (3.7) gives I in amperes when m is in grams per second, t in seconds, D in centimeters-squared per second, and C is moles per metre. It describes the current during the lifetime of each drop, hence is inherently discontinuous with respect to both time and voltage. The equation must be applied separately for each drop, using the newly incremented voltage. This equation agrees with experimental results within a few percent.

Two limiting cases need to be treated: (1) where E is much more negative than $E_{1/2}$, so that we are situated on the plateau of the polarogram, and (2) the case where only the last instant of the lifetime of each drop is considered, and time t is replaced by a constant τ *called the drop time.*

At a sufficiently negative voltage, the value of θ becomes small, since it aproximates the ratio C^S_{OX}/C^S_{RED}, which itself becomes very small. In this case Eq. (3.7) reduces to the *Ilkovic equation [9]*

$$I = 7.06 \times 10^3 \cdot nm^{2/3}\, t^{1/6}\, D^{1/2}_{OX}\, C^*_{OX} \tag{3.8}$$

The *diffusion current*, i_d, is the net current flowing just prior to the end of the drop-life, as shown is Figure 3.2. The drop time, τ, can be substituted for t to give:

$$i_d = 7.06 \times 10^3 \cdot nm^{2/3}\, t^{1/6}\, D^{1/2}_{OX}\, C^*_{OX} \tag{3.9}$$

The numerical multiplier is often given as 706, which gives the current in microamps if m is taken in milligrams per second and the concentration in millimoles per liter. Older instrumentation, now obsolete, used a highly damped recording mechanism that gave, in effect, the *average* current. To account

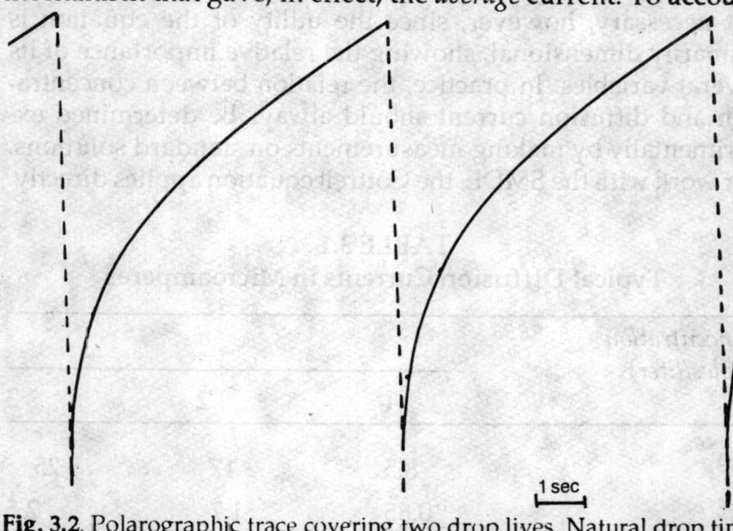

Fig. 3.2. Polarographic trace covering two drop lives. Natural drop time, about 6.5 s; paper speed, 1 cm/s. (Recorded on a Sargent-Welch Polarograph).

for this, Eq. (3.10) must be integrated and divided by the drop time, to give.

$$\bar{I} = 605\ nm^{2/3}\tau^{1/6} D_{OX}^{1/2}\ C_{OX}^{*} \qquad (3.10\ a)$$

These equations demonstrate a linear relation between the diffusion current and the concentration, a property which is exploited in analytical measurements.

The order of magnitude of the diffusion current as a function of $n\ C_{OX}^{*}$ may be seen by taking as typical values, $m = 0.002\ g\ s^{-1}$, $\tau = 4s$, and $D_{OX} = 6 \times 10^{-6} cm^2\ s^{-1}$, giving the results shown in Table 3.1.

It is noteworthy that the temperature does not appear explicitly in the Cottrell and Ilkovic equations. It does, however, have an effect, primarily through the temperature dependence of the diffusion coefficient and of the physical properties of mercury. The net effect on the diffusion current is to increase its value by 1 or 2 per cent per degree rise of temperature in the absence of kinetic complications. For this reason, it is advisable to control the temperature of a polarographic cell to within about 0.5° C if quantitative work is contemplated.

The Ilkovic equation can be made more precise by the addition of one or more terms. For analytical purposes this is not necessary, however, since the utility of the equation is primarily dimensional, showing the relative importance of its several variables. In practice, the relation between concentration and diffusion current should always be determined experimentally by making measurements on standard solutions. For work with the SMDE, the Cottrell equation applies directly.

TABLE 3.1
Typical Diffusion Currents in Microamperes

Concentration (moles/liter)	*n*		
	1	2	3
10^{-3}	8.5	17	25
10^{-4}	0.85	1.7	2.5
10^{-5}	0.085	0.17	0.25

The Heyrovsky-Ilkovic Equation

We will now derive an equation for the shape of the curve of faradaic current plotted against applied potential, which is approximated by the sampled polarogram. This is tantamount to taking the locus of all point for which $t = \tau$. Note that now the current is independent of time.

Substitution of the value of i_d from Eq. (3.9) into Eq. (3.7) gives (with replace ment of t by τ):

$$l = i_d \left(\frac{1}{1+\theta}\right) \tag{3.12}$$

which can be solved for θ to give:

$$\theta = \frac{i_d - l}{I} \tag{3.13}$$

from Eq. (3.3), by taking logarithms, we can write:

$$E = E_{1/2} + \frac{RT}{nF}\, ln \left(\frac{i_d - I}{I}\right) \tag{3.14}$$

This equation, first derived by Heyrovsky and Ilkovic, gives the shape of the current-voltage curve, which is centro-symmetric about $E_{1/2}$. The relationship between E_{id} and l is evident is Figure 3.3.

The half-wave potential, $E_{1/2}$, has two definition. Mathematically, it is defined as

$$E_{1/2} = E^{o'} + \frac{RT}{nF} \ln \left(\frac{D_{RED}}{D_{OX}}\right)^{1/2} \tag{3.15}$$

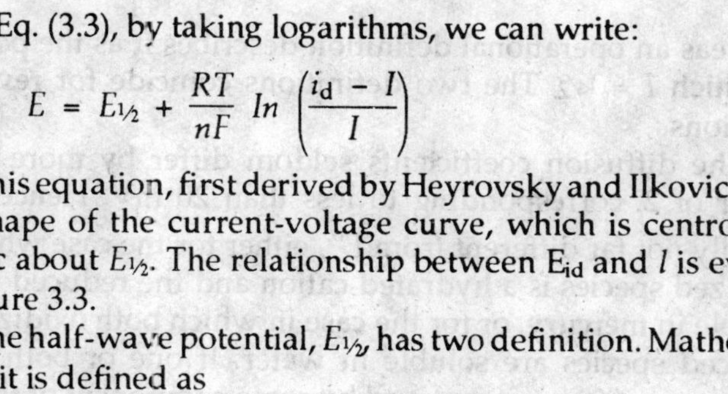

Fig. 3.3. Idealized polarogram, illustrating the relations between l and i_d.

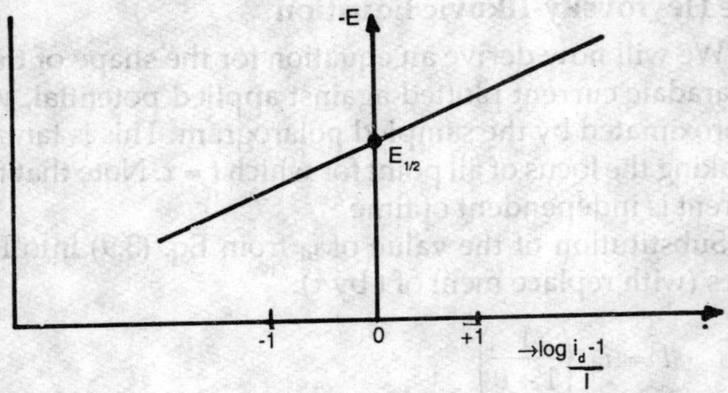

Fig 3.4. Plot of potential against the negative of log $\left[\frac{i_d - I}{I}\right]$ showing how to determine $E_{1/2}$.

whereas an operational definition describes it as the potential at which $I = i_d/2$. The two definitions coincide for reversible reactions.

The diffusion coefficients seldom differ by more than a factor of 2, corresponding to less than 20 mV. Hence, $E_{1/2}$ is usually not far different from $E^{o\prime}$, either for the case where the oxidized species is a hydrated cation and the reduced form is soluble in mercury, or for the case in which both oxidized and reduced species are soluble in water. If one or both water-soluble species are complexed by some component of the solution, or if kinetic complications appear, then $E_{1/2}$ may differ considerably from $E^{o\prime}$.

If the function $-\log [i_d - I)/1]$ is plotted against $-E$, a straight line should result with slope equal to 0.0592/n volts (at 25°C), crossing zero at $E_{1/2}$ (Figure (3.4). This provides a convenient method of determining both n and $E_{1/2}$ for reversible systems.

The diffusion current (i_d) is a measure of the rate of diffusion of the electro-active species and is directly proportional to the concentration of the substance under assay in the test solution. The limiting current, as is shown by the Figure 3.5, is the sum of the residual current and the diffusion current. The residual is subtracted from the limiting current to give the wave height. The linear relationship between the diffusion current and the concentration of electro-active species is shown by the Ilkovic equation,

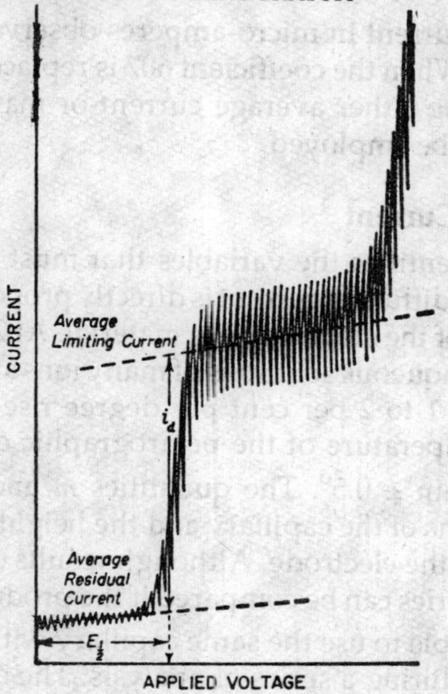

Fig 3.5. Typical Polarogram showing change in current flow with potential applied to the Dropping Mercury Electrode.

$$i_d = 607n\, D^{1/2}\; Cm^{2/3}\, t^{1/6},$$

in which i_d is the average current in microamperes, n is the number of electrons required per molecule of electro-active substance, D is its diffusion coefficient in square centimeters per second, C is the concentration in millimoles per litre, m is the rate of mercury flow from the dropping mercury electrode in mg per second, and t is the drop time is seconds.

According to the Ilkovic equation, the diffusion current increases from a small value as the drop begins to form to a maximum value as the drop falls. The diffusion current then falls abruptly to the low value and again increases in the same manner. By the use of a suitable recorder to measure the current, a saw-toothed record is obtained. For Instruments using galvanometers to measure the current or pen recorders adjusted to simulate galvanometer recording, the saw-toothed variations correspond to oscillations about the average current; thus the midpoints of the oscillations represent the average current. For pen recorders that are sufficiently rapid to follow the current during the second half of the drop life, but are too slow to respond to rapid current changes and do not show the current falling to a value near zero, the maximum of the oscillations should be determined as a measure of the current. For polarograms obtained in this manner, the i_d given by the Ilkovic equation is the maximum current in micro-amperes observed during the life of the drop. When the coefficient 607 is replaced by 708. An instrument using either average current or maximum current readings may be employed.

Control of the Diffusion Current

The Ilkovic equation identifies the variables that must be controlled to ensure that the diffusion current is directly proportional to the concentration of the electro-active material. At 25° the diffusion coefficient for aqueous solutions of many ions and organic molecules increase 1 to 2 per cent per degree rise in temperature. Thus, the temperature of the polarographic cell must be controlled to within ± 0.5°. The quantities m and t depend upon the dimensions of the capillary and the height of the mercury column above the electrode. Although results obtained with different capillaries can be compared if the product $m^{2/3}\,t^{1/6}$ is known, it is advisable to use the same capillary with a constant head of mercury during a series of analysis. The diffusion current is proportional to the square root of the height of the mercury column. A mercury reservoir with a diameter greater than 4 cm prevents any significant drop in the mercury level during a series of runs.

The capillary for the dropping mercury electrode has a bore of approximately 0.04 mm and a length of 6 cm. to 8 cm. The height of the mercury column, measured from the tip of the capillary to the top of the mercury pool, ranges from 40 cm to 80 cm. The exact length of the capillary and the height of the mercury column are adjusted to give a drop-time of between 3 and 5 seconds at open circuit with the capillary immersed in the test solution.

Equipment is available that allows controlled drop-times of fractions of a second. As detail within a polarogram is related to the number of drops delivered during a given potential change, such short drop times allow rapid recording of the polarogram.

The current flowing through the test solution during the recording of a polarogram is in the micro-ampere range. Thus, the current flow produces negligible changes in the test solution and several polarograms can be run on the same test solution without significant differences.

Half-wave Potential

The half-wave potential ($E_{1/2}$) occurs at the point on the polarogram onehalf the distance between the residual current and the limiting current plateau. This potential is characteristic of the electro-active species and is independent of its concentration or the capillary used to obtain the wave. It is dependent upon the solution composition and may change with variations in the pH or in the solvent system or with the addition of complexing agents. The half-wave potential thus serves as a criterion for the qualitative identification of a substance.

The potential of the dropping mercury electrode is equal to the applied voltage versus the reference electrode after correction for the *iR* drop (that potential needed to pass the current, *i*, through the solution with a resistance *R*). It is especially important to make this correction for non-aqueous solutions, which ordinarily possess high resistance, if an accurate potential for the D.M.E. is needed. Correction of the half-wave potential is not required for quantitative analysis. Unless otherwise indicated, it is to be understood that potentials represent measurements made against the saturated calomel electrode.

Removal of Dissolved Oxygen

Inasmuch as oxygen is reduced at the D.M.E. in two steps, first to dydrogen peroxide and then to water, it interferes where palarograms are to be made at potentials more negative than about 0 volt versus S.C.E., and must be removed. This may be accomplished by bubbling oxygen-free nitrogen through the solution for 10 to 15 minutes immediately before recording the wave, the nitrogen first having been "conditioned" to minimize changes due to evaporation, by being passed through a separate portion of the solution.

Sodium bisulphite may be added to remove oxygen, provided the reagent does not react with other components of the solution. It is necessary that the solution be quiet and vibration-free during the time the wave is recorded, to ensure that the current is diffusion-controlled. Therefore, the nitrogen aeration should be stopped and the gas be directed to flow over the surface of the solution before a polarogram is recorded.

Measurement of the Wave Height

A polarogram is used quantitatively by measurement of the vertical distance between the extrapolated line representing the residual current and that representing the average current. This height at the half-wave potential, $E^{1/2}$ is a graphical expression of the diffusion current *id*, as shown in the Figure 3.5.

Procedure

(*Note* - Mercury vapour is polsonous.) Work in a well-ventilated laboratory, taking care to clean up any spilled mercury.

Transfer a portion of the final dilution of the substance being assayed to a suitable polarographic cell immersed in a water bath regulated to 25 ± 0.5°. Pass a stream of nitrogen through the solution for 10 to 15 minutes to remove dissolved oxygen. Start the mercury dropping from the capillary, insert the capillary into the test solution, and adjust the height of the mercury reservoir. Switch the flow of nitrogen to pass over the surface of the solution, and record the polarogram over the potential range indicated in the individual monograph, using the appropriate recorder or galvanometer sensitivity to give a suitable wave. Measure the height of the wave, and compare this with a calibration curve obtained with the appropriate Reference Standard under the same conditions.

Dependence on Mercury Pressure

In DME the product of flow rate (*m*) and drop time (*t*) are equal to the weight of a single drop. At a constant potential, the drop weight would be constant ($w = m \times ti$). By substituting for $\frac{W}{m}$ for *t* in Iikovic equation we find that a fixed concentration of the depolariser i.e.

$$i_d = k^1 \sqrt{m}$$

Since the flow rate is proportional to the leight of the mercury column (*h*), the diffusion current is also proportional to the square root of *h*.

$$i_d\, k^1 \sqrt{h}$$

The linear dependence of i_d with $\sqrt{h}$ is an important criterion for the occurrence of purely diffusion current. This relationship has an important application in practical polarography. It has yielded a simple and commonly used test to establish the diffusive character of any limiting current. For this purpose limiting current at a fixed potential is measured at different heights of mercury column and plotled against $\sqrt{h}$. In case the current is diffusion controlled, the plot yields a straight line passing through the origin. To carry out precise work a correction for back pressure of mercury should the applied while computing the height of the column. For this correction the following term is used:

$$P_{back} = \frac{3.1}{m^{1/3}\,t^{1/3}}\ cm$$

The term should be deducted from the measured height of the mercury column.

i.e., $h_{eff} = h$ - correction term.

Influence of Temperature

It has been observed that the diffusion coefficient of a species in solution increases with increase in temperature. It results in an increase in diffusion current with increase in temperature. The *temperature coefficient* or the *percentage increase in the diffusion current per degree rise in temperature*, can be derived from Iikovic equation; It lies near about 1.6 per cent. While doing analytical work. The temperature must be maintained within ± 0.5°C in order to ensure that the error does not exceed 1%. Thus all solution including the reference electrode must be kept at the same temperature. The dependence of limiting current on temperature in an important test for characterising the diffusion current.

Influence of Supporting Electrolyte

Any changes in the composition of supporting electrolyte brings about changes in diffusion current. The influence is quite evident in those cases where some constituent of the supporting electrolyte co-ordinates with the metal ions under investigations. When ligands of larger size (such as tartarates, citrates etc.) are present, the diffusion current resulting from the metal ion tends to decrease because of the formation of bulky complex ions.

However, moderate changes in the concentrations of supporting electrolyte do not make any appreciable change in the diffusion current values. In general there is a decrease in value with increase in the concentration of supporting electrolyte. This is because of an increase in the viscosity of the medium. In most of the cases the diffusion current constants are larger in chlorides than in sulphates or nitrate media.

Limiting Current Measurements

We have already seen that the current observed in polarogram consists of diffusion current, changing current etc. The impurities present in the medium also make their contribution. It is possible to find the true value of diffusion current only if we make corrections for residual current etc. At lower concentration of depolariser, the magnitude of the residual current becomes significant in the wave height.

When DME is used as the indicator electrode, the polarographic wave does not appear as a continuous line but displays oscillations which are dependent on the magnitude of the current, drop time and degree of damping. The mean of oscillations is taken for measuring the magnitude of the current. Normally, a line is drawn through the middle portion of the oscillations, for this purpose.

The correction for the residual current can be made by either of the following two modes:

(i) A current voltage wire is drawn in presence of supporting electrolyte alone (residual current). The values of residual current at different potentials are substracted algebraically from the corresponding values in the polarogram of the depolarizer. The resulting curve may be considered as true or corrected current voltage curve of the depolariser.

(ii) The second mode is based on graphical extrapolation (Fig 3.6) which is based on assumption that the residual current increases linearly with increase in potential. In this the residual portion of wave is extrapolated (Fig 3.6 line AC) while limiting current portion is drawn backwards (Fig 3.6 line BD). The line FG that passes between the middle points of two lines will cut the curve at $E_{1/2}$. The wave height is measured as the vertical line *KH* passing through $E_{1/2}$. This method is useful if there are several waves in the polarogram. (Fig 3.7).

(iv) Adsorption Currents

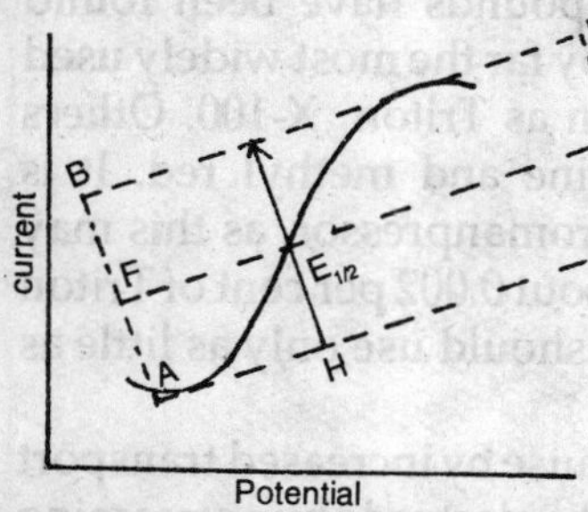

Fig 3.6. Graphical method of correcting residual current.

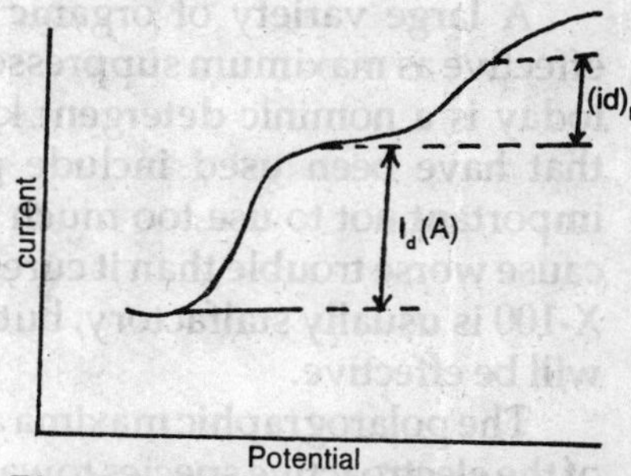

Fig 3.7. Multiple polarographic wave.

While working with large organic molecules, sometimes it has been observed that polarographic current is governed by the surface area of the electrode. This is due to the fact that many electroactive species, or their products from the electrode reaction, get adsorbed on the electrode surface and this adsorption phenomenon greatly influences the electrochamical reaction.

(v) Kinetic Current

The limiting current is also affected by the rate of non-electrode reaction and is known as *kinetic current*. This type of current results when oxidised or reduced form of electroactive species is in valued in chemical equilibrium with other species.

Study of kinetic waves provide useful information about reactions which is difficult to obtain otherwise.

(vi) Catalytic Current

In same cases very high current is observed is polarographic studies when trace of a substance is added. The magnitude of current is so much high that it cannot be expected due to normal electron-transfer process. Such a current, known as *catalytic current* is caused due to the presence of a substance in solution that acts as a catalyst during reduction or oxidation of a depolariger. Catalytic evolution of hydrogen causes the most commonly observed catalytic currents.

Polarographic Maxima

In a typical DC polarographic experiment, a small amount of a surface active agent is frequently needed. This it to prevent the formation of distortions called *maxima*.

A large variety of organic compounds have been found effective as maximum suppressors. By far the most widely used today is a nominic detergent known as Triton X-100. Others that have been used include gelatine and methyl red. It is important not to use too much sup romanpressor, as this may cause worse trouble than it cures. About 0.002 per cent of Triton X-100 is usually staifactory, but one should use only as little as will be effective.

The polarographic maxima are cause by increased transport of the electroactive species towards the electrode by a streaming motion of the solution.

Typical maxima obtained in polarograms are shown in Fig. 3.8.

The maxima of first type (curve A Fig 3.8) appear as sharp peaks with in a narrow potential range and are more pronounced at lower concentration of the supporting electrolyte.

The maxima of the second type (curve B. Fig. 3.8) appear over a longer potential and are rounded in appearance. These generally occur at higher concentration of electrolyte and are pronounced at high pressure of mercury they disappear at low pressure of mercury. This type of maxima are called non-streaming maxima.

A study of lowering of polarographic maxima is a useful mathod to estimate surface activity. Traces of surface tants can be estimated in solutions. Purity of water and natural products such as honey can be monitored by observing the presence or absence of naturally occurring surface active agents. Influencing the polarographic maxima can act as a sensitive method of quantifying many materials.

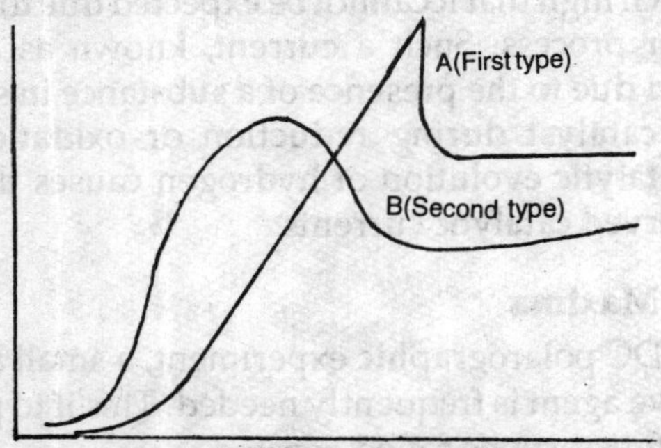

Fig 3.8. Polarographic Maxima.

4

Reversible Systems

Reversible Reaction : Product R is a Soluble Species

Consider the reversible electrode reaction $0_{(oxi)} + ne \rightleftarrows R_{RED}$, where the product

R is a species soluble in the solution.

(i.e., $Cr^{3+} + e \rightleftarrows Cr^{2+}$). The N crust relationship is

$$E_{O/R} = E_{0'_{O/R}} + \frac{0.0592}{n} \log \left(\frac{C^S_{OX}}{C^S_{RED}} \right) \tag{4.1}$$

where the concentrations (superscript *s*) are taken as surface values on either side of the interface, the reduced metal being on the mercury side.

Let us consider for purposes of illustration a system such as Zn^{++}/Zn, for which E^o = -1.00 V (vs. SCE), and for which n=2. We can calculate the ratio of concentrations C^S_{OX}/C^S_{RED} that must exist at the electrode surface for any applied potential *E* [Eq. (4.1]; these are presented in Table 4.1 for a series of potentials. We can see clearly from these figures that when the potential is further removed than about 100 mV from E^o, the ratio is very large in one direction and vanishingly small in the other. This means that for the less negative potentials, essentially all the electroactive solute is in the oxidized form (not surprising, since we started with the substance in the state). At the other extreme, there is essentially no solute in the oxidized state at the surface of the electrode.

TABLE 4.1

Necessary Ratio of Concentrationa $C^{S\,++}_{Zn}/C^{S}_{Zn}(Hg)$ at Electrode Surface for Various Applied Potentials (E)

E	*E-E°′*	$C^{S\,++}_{Zn}/C^{S}_{Zn}(Hg)$
0	1.00	6.08×10^{33}
–0.50	0.50	7.80×10^{16}
–0.90	0.10	2.39×10^{3}
–0.95	0.05	4.89×10^{1}
–0.99	0.01	2.18×10^{0}
–1.00	0.00	1.00
–1.01	–0.01	4.59×10^{-1}
–1.05	–0.05	2.05×10^{-2}
–1.10	–0.10	4.18×10^{-4}
–1.50	–0.50	1.28×10^{-17}
–2.00	–1.00	1.64×10^{-34}

The ratio of aqueous zinc ions to zinc atoms dissolved in mercury (as given by the table) determines the current that flows through the cell in the following way. At potentials more positive than about + 100 mV relative to $E^{o'}$, the ratio is large, but here the mercury contains no zinc, so the requirement is already fulfilled and very little faradaic curent is needed. On the other hand, as the potential becomes more negative, the ratio calls for a *larger* concentration of inc in the amalgam, and so a large current migh be expected. Here, however, the current is limited by the rate of diffusion of Zn^{++} ions from the bulk of the solution to the electrode surface, and a plateau in the current-voltage curve results. At intermediate potentials, near $E^{o'}$, the current is determined by diffusion across a lesser gradient and has an intermediate value.

The preceding analysis is based on the assumption that at all times the rate of the electrode reaction is limited only by diffusion. This is equivalent to assuming that the electrochemical system is *reversible*.

In the case of an irreversible system, the shape of the curve will be modified somewhat, though the initial and final currents will not differ from the reversible case. At positive potentials, the curves coincide because so little chemical work is called for that the current required is easily available. At negative potentials, they again coincide because here sufficient overvoltage has been applied so that the electron-transfer reaction proceeds rapidly. In between these extremes, however, diffusion can supply reducible material to the surface faster than the electrons can be made available to react with it, and the curve becomes unsymmetrically drawn out, as sketched in Figure 4.1.

As noted previously, a large concentration of an inert *supporting electrolyte* must be present in addition to the electroactive species under study. This has several important effects. It increases the conductivity of the solution, hence diminishing *IR* voltage drops; it makes certain that the double layer is established in a reproducible manner; it maintains a constant ionic strength and thereby ensures that the activity coefficient of the species under study does not change as the charge-transfer reaction proceeds; and it swamps out the migration current that otherwise would be a variable perturbing effect.

If no supporting electrolyte were present, migration would transport a considerable fraction of the ionic active species across the solution to the electrode. This woud change the concentration profiles as defined by diffusion and thus alter the current. Such a change is not objectionable *per se,* since the current is often increased as a result, but it has the fundamental disadvantage that it is affected by inert species such as Na^+ or

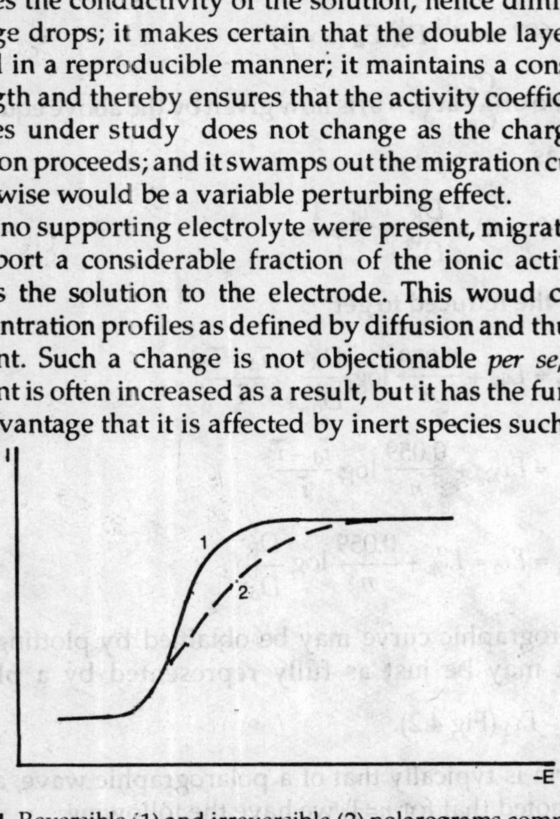

Fig. 4.1. Reversible (1) and irreversible (2) polarograms compared.

SO_4^- which would complete in carrying the current. Consequently, such species would constitute interferences. This mode of operation is generally not useful for analytical purposes.

In a reversible system if $E_{o/R}$ is the applied potential and C_o and C_R are the concentrations [0] and [R] at the electrode surface. The Iiokovic relationship yields

$$\bar{i} = 607\ nm^{2/3}t^{1/6}D_o^{1/2}\{C_o - C_o(x=0)\}$$

where $C_o(x = 0)$ is slightly less than C_o at an applied potential slightly more negative than the equilibrium potential for the original solution. This leads to

$$\bar{i} = \bar{i}_d - 607\ nm^{2/3}t^{1/6}D_o^{1/2}C_o(x=0) \quad ....(i)$$

since $C_o - C_o(x=0) = C_R(x=0)$, we have

$$\bar{i} = 607\ nm\ t^{1/6}D_R^{1/2}C_R(x=0) \quad ....(ii)$$

The ratio of $\frac{C_o}{C_R}$ at (x = 0) is now given by the above equatians (i) and (ii) as

$$\frac{C_o}{C_R}(x=0) = \frac{D_R^{1/2}}{D_o^{1/2}} \cdot \frac{\bar{i}_d - 1}{\bar{i}}$$

It can then the reduced to get

$$E_{o/R} = E^o_{o/R} + \frac{0.059}{A}\log\frac{D_R^{1/2}}{D_o^{1/2}} \cdot \frac{\bar{i}_d - \bar{i}}{\bar{i}}$$

$$= E_{1/2\,o/R} + \frac{0.059}{n}\log\frac{\bar{i}_d - \bar{i}}{\bar{i}}$$

where $E_{1/2\,o/R} = E_{1/2} = E^o_{o/R} + \frac{0.059}{n}\log\frac{D_R^{1/2}}{D_o^{1/2}}$

A polarographic curve may be obtained by plotting $\bar{i}$ Vs $E_{o/R}$, but it may be just as fully represented by a plot of $\frac{\bar{i}}{\bar{i}_d}$ Vs $E_{o/R} - E_{1/2}$ (Fig 4.2).

The form is typically that of a polarographic wave, and it should be noted that for n=1, we have the following:

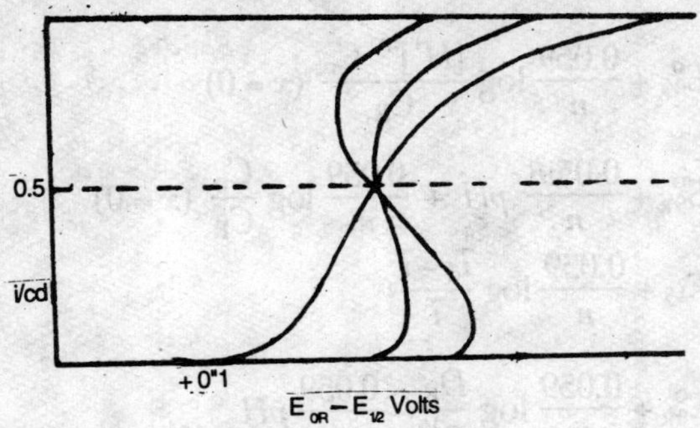

Fig 4.2. The polarographic curve or wave.

(i) When $E_{O/R}$ is about 0.1 V less negative than $E_{1/2}$, $\bar{i}$ is about 0.02 $\bar{i}d$ or much less than $\bar{i}_d$.

(ii) When $E_{O/R}$ is about 0.1 V more negative than $E_{1/2}$, $\bar{i}$ is about 0.98 $\bar{i}d$, or approximately equal to $\bar{i}_d$.

(iii) When $E_{O/R}$ is identical to $E_{1/2}$, $\bar{i}$ is exactly one half $\bar{i}_d$.

The plot of $\bar{i}$ Vs $E_{O/R}$ yields the same wave form, the plateau of wave then corresponds to the condition $\bar{i} = \bar{i}_d$.

The value of $E_{1/2}$ is known as *half wave potential* for the system and is independent of concentration for most reversible reaction. For such reactions the value of $E_{1/2}$ will be constant for a given reaction in a specific solution medium and can serve to identify species 0.

For reversible reactions in general, a plot of $E_{O/R}$, the applied cathodic potential Vs log $\left(\frac{\bar{i}_d - \bar{i}}{\bar{i}}\right)$ will yield a straight line from which $E_{1/2}$ may be determined as the intercept with the $E_{1/2}$ axis for log $\left(\frac{\bar{i}d - \bar{i}}{\bar{i}}\right) = 0$, will n being determined from the slope of the line.

Reversible reactions : Hydrogen ion Participates in the Electrode Reaction.

The general reaction in this case at DME is

$$O_{+} m \ H + ne \rightleftarrows R + W H_2O,$$

where the degree of reaction completion depends upon $[H^+]$ and where R is a solution-soluble species.

$$E_{O/R} = E^{o}_{O/R} + \frac{0.059}{n} \log \frac{[H^{+}]^{m}\ C_{o}}{C_{R}} \quad (x = 0)$$

$$= E^{o}_{O/R} + \frac{0.05m}{n}\ pH + \frac{0.059}{n} \log \frac{C_{o}}{C_{R}} \quad (x = 0)$$

$$= E_{1/2} + \frac{0.059}{n} \log \frac{\bar{i}_{d} - \bar{i}}{\bar{i}}$$

where $$E_{1/2} = E^{o}_{O/R} + \frac{0.059}{n} \log \frac{D_{R}^{1/2}}{D_{o}^{1/2}} - \frac{0.059}{n}\ pH$$

$E_{1/2}$ for a specific pH can be determined in the usual may be plotting $E_{O/R}$ Vs log $\left(\frac{\bar{i}_{d} - \bar{i}}{i}\right)$, with *n* being determind from straight line slope, $E_{1/2}$ may be determined for different solution values of pH. The plot of $E_{1/2}$ Vs pH is a straight line from which values of *m*, *n* can be found.

Reversible Reaction : Metal Ion Reduction with Amalgam Formation

In this case *O* is represented by metal ion m^{n+} and *R* by the metal *M*. The metal is capable of amalgam formation with mercury and when the reaction.

$m^{n+} + ne \rightleftarrows M$ takes place at the DME surface, the metal diffuses into the drop, D_R being the diffusion coefficient for this process. The following equation holds good for reaction of this type.

$$E_{O/R} = E_{1/2},a + \frac{0.059}{n} \log \frac{\bar{i}_{d} - \bar{i}}{\bar{i}}$$

where

$$E_{1/2},a = E^{o}_{a}, M^{n+/M} + \frac{0.059}{n} \log \frac{D^{M^{1/2}}}{(D_{m}^{n+})^{1/2}}.$$

The value of E^{o} may be significantly different from $E_{1/2}$. As in previous case $E_{1/2}$, *a* and *n* may be determined for reversible reaction by plotting $E_{O/R}$ Vs log $\left\{\frac{\bar{i}_{d} - \bar{i}}{i}\right\}$

Reversible Reaction : Metal Complex Ion Reduction with Amalgam Formation.

Here we have the metal complex ion $M(L)_x^{(n-xp)}$ of the ligand L^{p-} dissociating $M(L)_x^{(n-xp)} = M^{n+} + x\,L^{p-}$ with M^{n+} being reduced to M at the DME with amalgam formation. $M(L)_x^{(n-xp)}$ must then diffuse to the electrode surface, we have

$$\frac{[M^{n+}]\,[L^{p-}]^x}{[M(L)_x^{n-px}]} = k\ inst$$

For the dissociation of metal complex ion

$$E_{o/R} = E^o_{a,\,o/R} + \frac{0.059}{n}\log\,[M^{n+}]$$

For the reduction of metal ion, where *o* and *R* represent M^{n+} and M, we have

$$E_{o/R} = E^o_{c,\,o/R} + \frac{0.059}{n}\log\left[M(L)_x^{(n-xp)}\right]$$

For the reduction of metal complex ion where O and R represent $M(L)_x^{(n-px)}$ and M the value of E^o_c is the standard reduction potential for the reduction of metal complex ion. Lingane and Von stackelberg have shown independently the relationship between E^a_a and E^o_c.

$$E^o_a + \frac{0.059}{m}\log\,[M^{n+}] = E^o_c + \frac{0.059}{n}\log\frac{[M^{n+}]\,[L^{p-}]^x}{K_{inst}}$$

so that

$$E^o_c = E^o_a - \frac{0.059}{n}\log\frac{[L^{p-}]^x}{K_{inst}}$$

The values of $E_{1/2},\,a$ and $E_{1/2},\,c$ will be given by

$$E_{1/2},\,a = E^o_a + \frac{0.059}{n}\log\frac{(D_m)^{1/2}}{(Dm^{n+})^{1/2}}$$

$$E_{1/2}\, c = E_c^o + \frac{0.059}{n} \log \frac{(D_M)^{1/2}}{\left[D_M(L)_x^{(n-px)}\right]^{1/2}}$$

For a general reversible reaction the above equation are valid, we can also write.

$$E_{1/2} c = E_a^o - \frac{0.059}{n} \log \frac{(L^{P-})^x}{K_{inst}} + \frac{0.059}{n} \log \frac{(DM)^{1/2}}{\left[D_M(L)^{(n-xp)}\right]^{\frac{1}{2}}}$$

$$= E_{1/2} a - \frac{0.059}{n} \log \frac{[L^{P-}]^x}{K_{inst}} + \frac{0.059}{n} \log \frac{(D_M n+)^{1/2}}{[D_M(L)x^{n-xp}]^{1/2}}$$

$$= E_{1/2} a - \frac{0.059}{n} \log (L_{p-}) + \frac{0.059}{n} \log k_{inst} +$$

$$\frac{0.059}{n} \log \frac{(D_M n+)^{1/2}}{\left[D_M(L)_x^{n-xp)}\right]^{1/2}}$$

Now that the magnıtude of $E_{1/2}$ depends on magnitude of k_{inst} and the concentration of the ligand. Where the metal complex ion is very stable, the value of $E_{1/2}, c$ will be very much more negative, than $E_{1/2}$, a difference of 1 to 2 *v.* more negative being more common for stable complexes. For reversable reactions, $E_{1/2}$, *a* and *n* may be determied in the usual manner from a plot of $E_{O/R}$ Vs log $\left\{\frac{\bar{i}_d - \bar{i}}{\bar{i}}\right\}$ for absence of the ligand.

Again for versible reaction involving the metal complex ion $M(L)_x^{n-np} + nc \rightleftarrows M + x L_1^{P-}$, we may determine the value of $E_{1/2}, c$ each of several solution of metal complex ion with different excess ligand concentration by plotting the applied potential $E_{O/R}$ Vs log $\left\{\frac{i_d - i}{i}\right\}$ for each of Hence solution. If we now plot the values of $E_{1/2}, c$ against log $[L^{P-}]$, we shall for revessible reactions obtain a straight line using the determined value of *n*, the slope of this line will permit evalution of *x*, the co-ordination number for the metal complex ion. For a fixed value of $[L^{P-}]$ using the determined value of *n*, *x* and $E_{1/2}$, *a*, the determined value of $E_{1/2}$, *c* for the value of $[L^{P-}]$ involved and assuming that

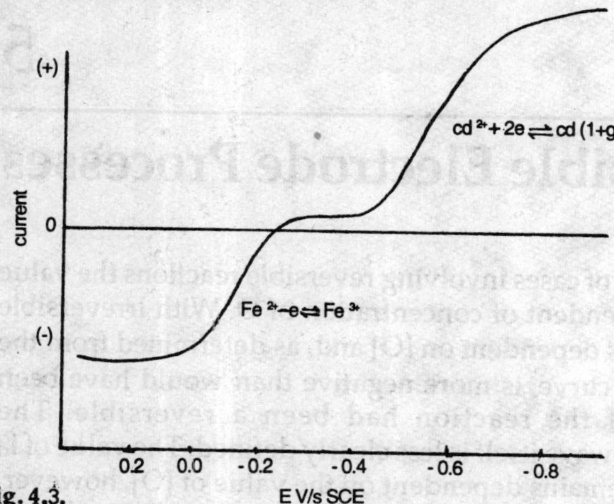

Fig. 4.3.

$(D_m^{n+})^{1/2} / \left[D_M(L)_x^{n-xp}\right]^{1/2}$ will opproximate unity. The value of K_{inst} for metal complex ion can be determined using the above equation.

It is important to note that where the values of $E_{1/2},1$ and $E_{1/2},2$ for two uncomplexed metal ion species M_1^{n+} and M_2^{n+} may be too close for polarographic separation and their subsequent identification, induced complex reaction in which are species is not complexed or yields a much less stable complex than the other can result in $E_{1/2},c,1$ and $E_{1/2},c,2$ values sufficiently separated to permit identification.

Anodic Wave : It is possible to obtain anodic polarographic waves where the DME functions as an anode and the electrode reaction involves $R - ne \rightleftarrows 0$. The limits in this area is the fact that applied potential values more positive than about 0.35 V *Vs* SCE, the ouidation of mercury takes place. It is also possible to obtain *anodic cathodic polarographic waves* where the electrode reaction (and of course the current flow) is reversed during the applied potential increase to more negative values.

Anodic-Cathodic waves may also be obtained where an anodic wave for species *R* (l) may be followed by a cathodic wave for species 0 (2). Fig 4.2 shows the polarogram for a solution containing Fe^{2+}/cd^{2+} and where the anodic wave for $Fe^{2+} - e \rightleftarrows Fe^{3+}$ is followed by the cathodic wave for

$$Cd^{2+} + 2e \rightleftarrows cd\ (Hg)$$

5

Irreversible Electrode Processes

In the majority of cases involving reversible reactions the value of $E_{1/2}$ is independent of concentration of O. With irreversible reactions, $E_{1/2}$ is dependent on [O] and, as determined from the polarographic curve, is more negative than would have been anticipated if the reaction had been a reversible. The polarographic wave itself is less clearly defined. The value of i_d is such cases remains dependent on the value of [O], however, so that irreversible reactions are also adaptable to the analytical applications of polarography.

Irreversible System

The equations presented in the preceding pages apply to ideal, reversible reactions. We now turn to irreversible systems for which some of the equations must be modified.

The basic distinction between the two classes lies in the magnitude of the heterogeneous rate constant, k^o, corresponding to the reaction :

$$OX + ne^- = RED$$

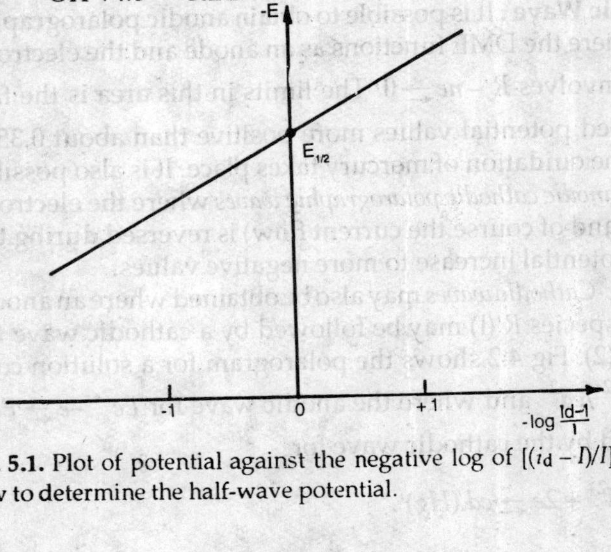

Fig. 5.1. Plot of potential against the negative log of $[(i_d - I)/I]$, showing how to determine the half-wave potential.

A convenient criterion for reversibility [11] is the relative values of the quantities $2(D/\pi\tau)^{1/2}$ and k^{o}, both of which have the dimensions of cm/s. The former can be considered to be a diffusion rate constant, k_d, while k^{o} is the charge transfer rate constant. We can state the conditions for reversibility as:

$k^{o} >> k_d$ (reversible)

$k^{o} << kd$ (irreversible)

In the case of polarography, we can take as typical values $D = 5 \times 10^{-6}\ cm^2 s^{-1}$, $\tau = 5\ s$, which gives for k_d a fairly constant value about $1 \times 10^{-3}\ cm \cdot s^{-1}$. As a rule, for a truly reversible reaction, $k^{o} > 2 \times 10^{-2}$ while a reaction for which $k^{o} < 2 \times 10^{-5}$ is considered to be totally irreversible.

In an electrochemical process, the tendency for a reaction to occur is determined by the applied potetial E, and therefore the reference potential must be specified. For present purposes, it is most convenient to take the formal potential E^{o}, as this reference. The following equations will be consistent with the standard.

The Cottrell equation is valid for both reversible and irreversible reactions because the potential step on which it is based extends sufficiently negative that all OX molecules reaching the electrode are immediately reduced, regardless of the magnitude of the rate constant. For a voltage step of arbitrary amplitude, the following relation can be obtained.

$$I = \frac{nFAD_{OX}^{1/2}}{(\pi\tau)^{1/2}\left[\exp\left\{\frac{nF}{RT}(E - E^{o\prime})\right\} + 1\right]} \cdot C_{OX}^{*}$$

in which it is assumed that $D_{OX} = D_{RED}$, so that $E_{1/2}$ can be replaced by E^{o}, when it reduces to Eq. already given for sufficiently negative potentials. This general relation, however, assumes reversibility, since it utilizes the Nernst equation in its derivation.

For an irreversible reaction, the comparable expression is

$$I = \frac{nFAk^{o}}{\exp\left\{\frac{\alpha nF}{RT}(E - E^{o\prime})\right\}} \cdot C_{OX}^{*}$$

where α is the transfer coefficient. Note the $(D_{OX}/\pi t)^{1/2}$ no longer appears, and k° takes its place, indicating that the proess is now under kinetic rather than diffusion control. For potentials sufficiently negative, the denominator becomes very small (for $E - E^{o'} = -1$ V, when $n = 2$, and α=0.5, the exponential is of the order of 10^{-17}), and this predicts an impossibly large current. The fact that the current does not really go this high is evidence that diffusion must take over control if the potential step is large enough.

Note that the Ilkovic equation, relatıng the diffusion current to concentration, is equally valid for reversible and irreversible processes, since the measurement of i_d is made under conditions that ensure diffusion control. Both cases are analytically useful.

The derivation leading to the Heyrovsky relation is no longer valid for an irreversible reaction, as one might expect from the difference in curve shapes. The derivation is too involved to justify repeating here. The final equation.

$$E = E_{1/2} + \frac{0.0235}{\alpha n} \ln\left(\frac{i_d - I}{I}\right)$$

where

$$E_{1/2} = E^{o'} + \frac{RT}{\alpha nF} \ln\left[k^o\left(\frac{12\tau}{7D_{OX}}\right)^{1/2}\right]$$

Note that $E_{1/2}$ is still the potential at which the current equals half the diffusion current, but its numerical significance is different. In this irreversible case, $E_{1/2}$ changes also with the drop time, τ.

Meites has suggested a method of testing for reversibility by measurement (on the polarogram) of $E_{1/4}$ and $E_{3/4}$, the potentials at which the corrected current is, respectively 1/4 and 3/4, of i_d. For a reversible reaction, one can show that :

$$E_{1/4} = E_{1/2} - \frac{0.0592}{n} \log (1/3)$$

and

$$E_{3/4} = E_{1/2} - \frac{0.0592}{n} \log 3$$

Substraction gives :

$$E_{3/4} - E_{1/4} = -\frac{0.0565}{n}$$

For an irreversible system, $E_{3/4} - E/4$ is generally more negative than this. This method can be used to determine n if the reaction is reversible, or as a test for irreversibility of n is known,

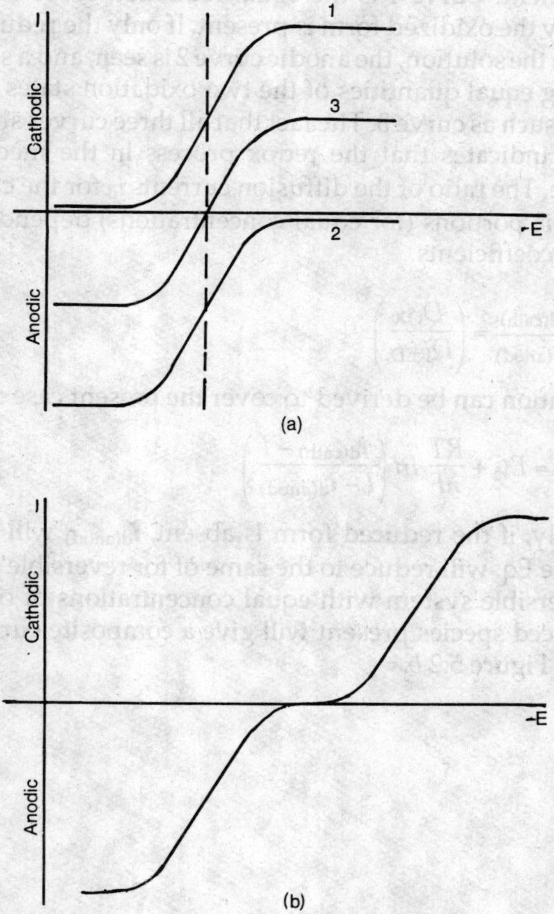

Fig. 5.2. Composite anodic and cathodic curves. (a) A reversible system, showing the pure components plus the mixture. (b) An irreversible system; the curve does not cross the zero-current axis abruptly, as in (a) but becomes tangent to it, because of the separation of half-wave potentials of the oxidized and reduced forms.

whereas the method of Figure 5.1 is only valid for reversible systems.

Anodic Oxidations

So far we have considered only cathodic reductions, but the DME can also act as an anode. Figure 5.2 shows the polarograms that result from a mixture of oxidized and reduced forms of the same element. Curve 1 is the usual reduction curve obtained when only the oxidized form is present. If only the reductant is present in the solution, the anodic curve 2 is seen, and a solution containing equal quantities of the two oxidation states gives a response such as curve 3. The fact that all three curves show the same $E_{½}$ indicates that the redox process in the medium is reversible. The ratio of the diffusion currents i_d for the cathodic and anodic portions (for equal concentrations) depend on the diffusion coefficients.

$$\frac{i_{d(cath)}}{i_{d(anod)}} = \left(\frac{D_{OX}}{D_{RED}}\right)^{½}$$

A relation can be derived to cover the present case :

$$E = E_{½} + \frac{RT}{nF}\, In \left(\frac{i_{d(cath)} - I}{I - i_{d(anod)}}\right)$$

Clearly, if the reduced form is absent, $i_{d(anod)}$ will be zero and above Eq. will reduce to the same of for reversible system. An irreversible system with equal concentrations of oxidized and reduced species present will give a composite curve such as that in Figure 5.2 *b*.

6

Pulse and Square-Wave Polarography

Many attempts have been made throughout the years to improve the sensitivity of the polarographic method. The most successful of these depend on some type of modulation, an approach that provides *two* qualitatively different parameters that can be measured, so that the effects of the changing current can be eliminated between them.

One of these methods is *pulse polarography,* in which a pulse of a few tens of millivolts is impressed on the staircase voltage toward the end of each drop-life. Two current measurements are taken, one just before the pulse, and another during the pulse, as shown in Figure 6-1. The *difference* between the current at B and that at A for each step, plotted against-*E*, gives a differential polaogram, essentially free from the effects of charging current. Pulse polarography and closely related methods are treated in this chapter.

Another technique, *AC polarography,* calls for the sup romanerimposition of a small amplitude AC voltage on the staircase ramp (AC modulation). The signal is taken as the AC component of the resulting current. Since alternating current is

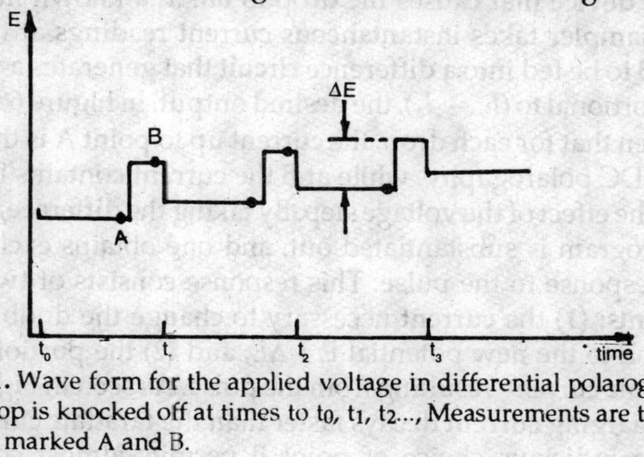

Fig. 6.1. Wave form for the applied voltage in differential polarography. The drop is knocked off at times to t_0, t_1, t_2..., Measurements are taken at points, marked A and B.

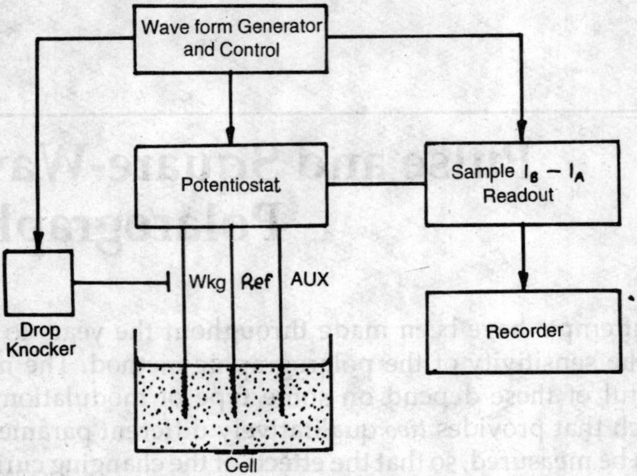

Fig. 6.2. Block diagram of a differential pulse polarograph.

characterized by two parameters, its amplitude and phase, careful attention to phase relations permits the determination of faradaic current free of changing current. This will be explored in the following chapter.

Both methods achieve a marked improvement of perhaps a thousandfold in sensitivity, compared to conventional (DC) polarography.

A Basic Experiment

A possible measurement scheme is shown in Figure 6.2. Just as in sampled DC polarography, the instrument must contain some device that causes the drop to fall at a known moment. The sampler takes instantaneous current readings at times A and B to be fed into a difference circuit that generates a voltage proportional to (I_b – I_a), the desired output. In Figure 6.3, it can be seen that for each drop, the current up to point A is the same as in DC polarogaphy, while at B the current contains in addition the effect of the voltage step. By taking the differece, the DC polarogram is substantiated out, and one obtains exclusively the response to the pulse. This response consists of two components: (1) the current necessary to change the double layer from E to the new potential E + ΔE, and (2) the portion of the faradaic current resulting from the pulse. As seen in Figure 6.3, the charging current decays faster than the faradaic current, so that a judicious choice of point B permits almost complete

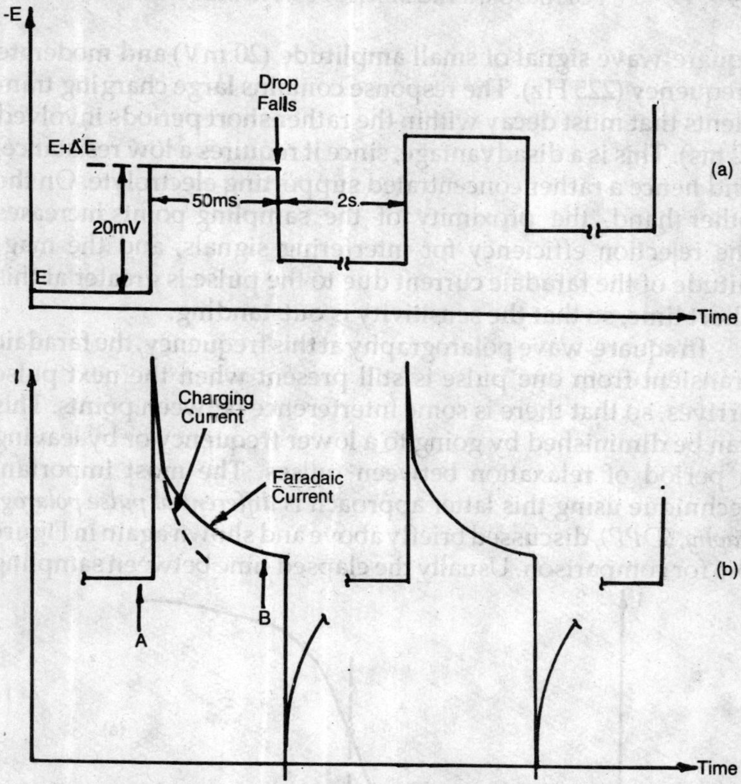

Fig. 6.3. Current response in differential pulse polarography.

elimination of the charging current. This represents the major strength of pulse polarography.

The fact that a difference is taken results in a derivative-like form for the signal, as illustrated in Figure 6.4. Note that the peak height of the signal is proportional to the concentration of the analyte, and that the baseline is nearly flat over a large range of concentrations.

Classification of Step Methods

A great amount of ingenuity has been devoted to designing various types of experiments involving voltage steps. Leaving aside certain older techniques that used imperfect sampling, and limiting ourselves to the DME, let us consider the most significant possibilities.

In Figure 6.5 is shown the first fully developed technique, the *square-wave polarography* of Barker, who used a symmetrical square-wave signal of small amplitude (20 mV) and moderate frequency (225 Hz). The response contains large charging transients that must decay within the rather short periods involved (2 ms). This is a disadvantage, since it requires a low resistance, and hence a rather concentrated supporting electrolyte. On the other hand, the proximity of the sampling points increases the rejection efficiency for interfering signals, and the magnitude of the faradaic current due to the pulse is greater at this short time, so that the sensitivity is outstanding.

In square-wave polarography at this frequency, the faradaic transient from one pulse is still present when the next pulse arrives, so that there is some interference between points. This can be diminished by going to a lower frequency or by leaving a period of relaxation between pulses. The most important technique using this latter approach is *differential pulse polarography, (DPP),* discussed briefly above and shown again in Figure 6-6 for comparison. Usually the elapsed time between sampling

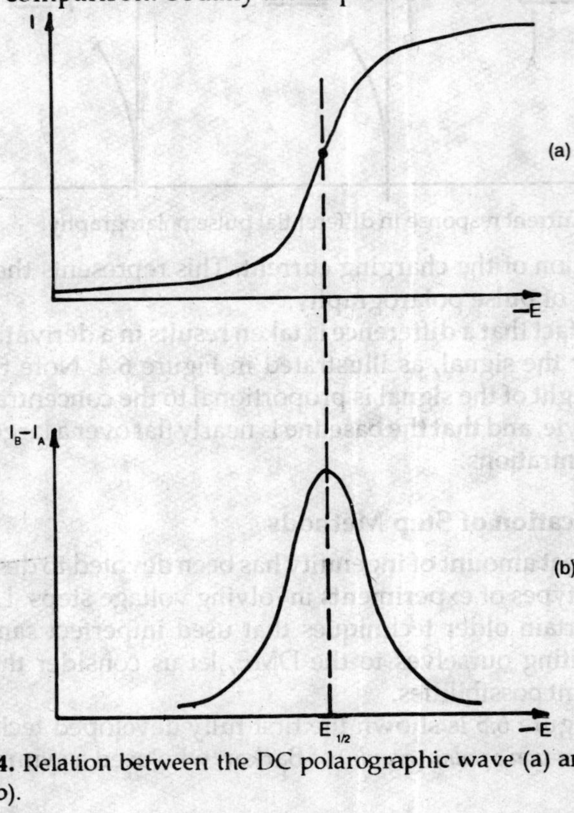

Fig. 6.4. Relation between the DC polarographic wave (a) and the DPP peak (b).

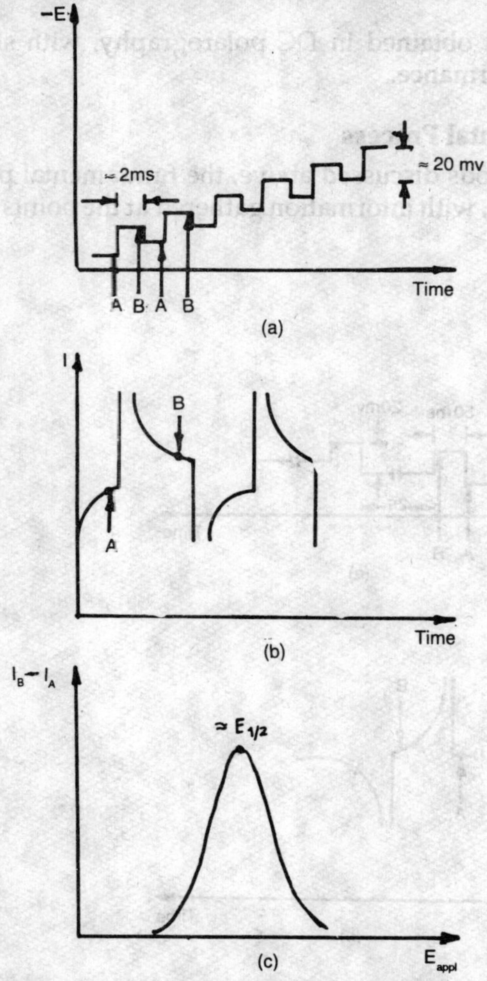

Fig. 6.5. Wave forms the square-wave polarography. (*a*) Applied potential (*b*) Current response, showing sampling points (*c*) Output.

points *A* and *B* is an order of magnitude larger than in square-wave polarography. This makes the elimination of charging-current effects much easier, but as the time base increases, the rejection of noise decreases, so that the two mthods have roughly the same sensitivity.

Also useful is the large-step procedure known as *normal pulse polarography*, illustrated in Figure 6.7. Either single-point sampling (at point *B*) or two point sampling (*B* – *A*) can be employed. In either case, the output gives a step or wave resembling that obtained in DC polarography, with slightly improved performance.

The Fundamental Process

In the methods discussed above, the fundamental process is a voltage step, with information gathered at the points A and

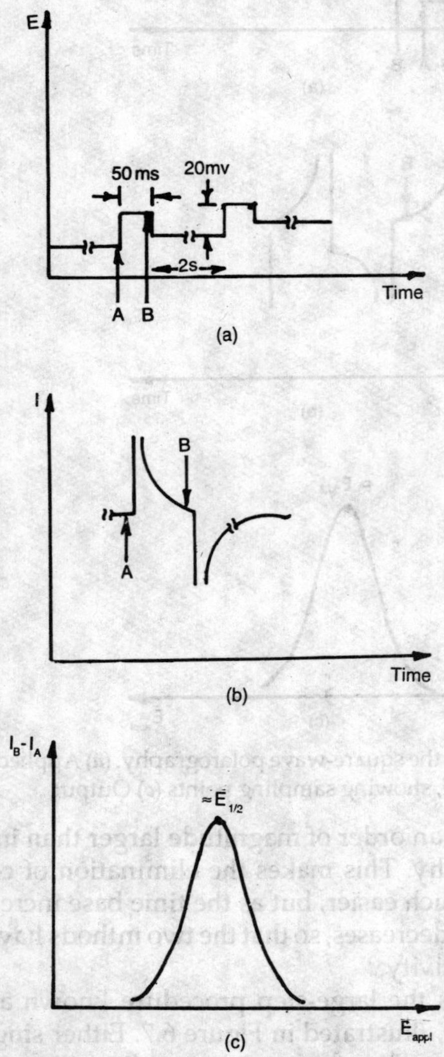

Fig. 6.6. Wave forms for DPP. (*a*) Applied potetial. (*b*) Current response, showing sampling points. (*c*) Output.

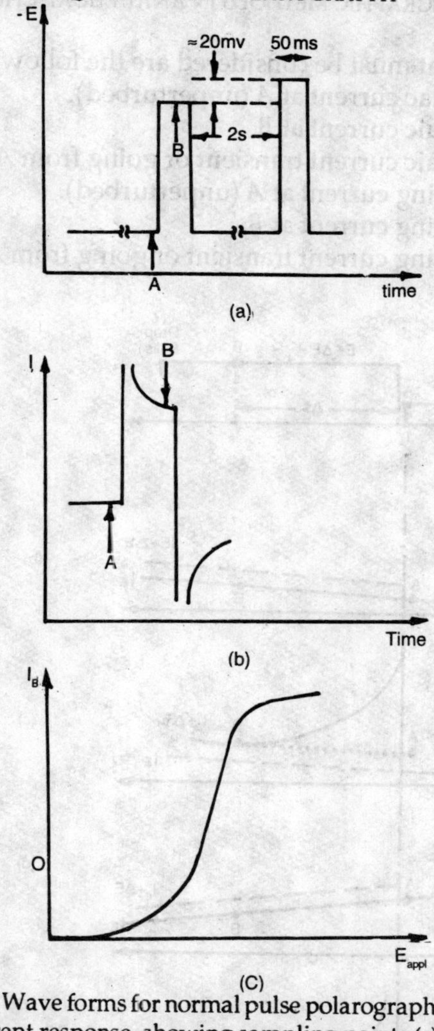

Fig. 6.7. Wave forms for normal pulse polarography. (*a*) Applied potential. (*b*) Current response, showing sampling points (*c*) Output.

B, before and after the transition. The separation between points is in the range of 2–100 ms, depending on the method.

As with DC polarography, the current contains two major components, faradaic and charging, but now the effect of the voltage step on both must be considered. We shall discuss the various currents separately, assuming no interaction between them. The change in the surface area of the mercury drop between *A* and *B* is normally small enough to be neglected, except as a residual effect.

The currents that must be considered are the following:

1. The faradiac current at *A* (unperturbed).
2. The faradaic current at *B*.
3. The faradaic current transient or going from *A* to *B*.
4. The charging current at *A* (unperturbed).
5. The charging current at *B*.
6. The charging current transient or going from *A* to *B*.

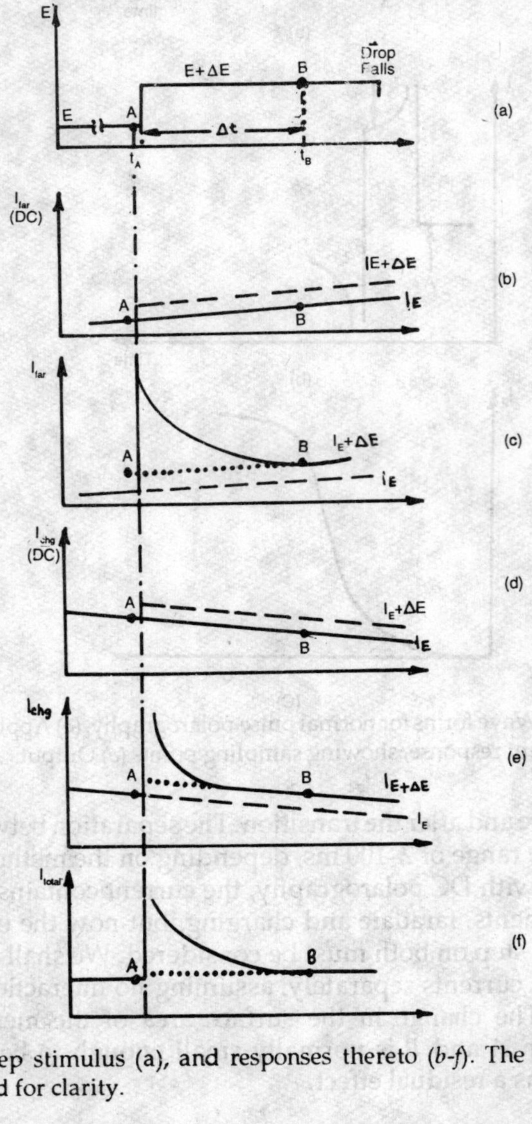

Fig. 6.8. Step stimulus (a), and responses thereto (*b-f*). The slopes are exaggerated for clarity.

The first component to be discussed is the faradaic current, represented in Figure 5.8 *b*. It can be through of as a small segment, length Δt, of an Ilkovic-type curves, $I = (\text{const}) \cdot t^{1/6}$, and can be approximated by a straight line with a small positive slope. The dashed line represents the current corresponding to $E + \Delta E$, which is the Ilkovic curve for the potential at point B. The separation between the two curve is greatest at $E_{1/2}$ and is zero in the diffusion-current region. This means that the effect of previous reductions is mostly eliminated, an important property of differential operation.

It is interesting to estimate the extent of this rejection. The only factor concerned is the effect of the small change of surface area between A and B. If we take as typical values the sampling time $t_a = 2$ s, and the interval $\Delta t = 0.05$ s, we can determine the fractional change caused by the voltage step:

$$\frac{I_B - I_A}{I_B} = \frac{(const) \cdot t_B^{1/6} - (const).\, t_A^{1/6}}{(const) \cdot t_B^{1/6}}$$

$$= 1 - \left(\frac{2.00}{2.05}\right)^{1/6} = 0.004$$

The effect is small, only 0.4 per cent of I_b, but not negligible. Note that as the voltage shifts from E to $E + \Delta E$, the faradaic current does not rise directly to the new value, but follows a transient curve, as shown in Figure 6.8 c. The transient decays with a $1/\sqrt{t}$ dependence, in accordance with the relation:

$$\Delta I = nFA \left(\frac{D_{OX}}{\pi \Delta t}\right)^{1/2} C_{OX}^{*} H(E_j, t_j) \qquad (6.2)$$

in which a reduction is assumed. The quantity ΔI, the difference between the currents at B and A, is the principal output of the polarograph. The function H depends in a rather complex way on the voltage E_j and time periods t_j of previous steps in the history of the drop. This function is unity if the pulse is large and comes from a region where there is no faradaic process to a point where diffusion currents prevail, as is true in normal pulse polarograpy. In this instance, the relation reduces to the Cottrell equation. In contrast, the function H plays an important role in square-wave operation.

The effect of the pulse on the charging current must also be considered. Its DC component is shown in Figure 6.8 *d*. This represents the principal factor that limits the sensitivity of DC polarography, since it becomes dominant for solutions of less than about 10^{-5} *M*. The charging current arises from the growth of the drop, and the need to endow the newly formed surface with its double layer. It depends, therefore, on the applied voltage, and has a different value of *E* and at $E + \Delta E$. This can be seen from the following argument, using the expression for the charge, *Q*, as a function of capacitance, C, the area, *A*, and the voltage, *E* :

$$Q = CAE \tag{6.3}$$

The charging current is then given by :

$$I_E = \frac{dQ}{dt} = CE\frac{dA}{dt} \tag{6.4}$$

and

$$I_{E+\Delta E} = C(E + \Delta E)\frac{dA}{dt} \tag{6.5}$$

Noting that the area increases at $t^{2/3}$, we obtain:

$$I_{chg} \; \alpha \; CEt^{-1/3} \tag{6.6}$$

a decreasing function of time. This indicates that longer drop-times are preferable in order to diminish the charging current contribution. In the time interval Δt between *A* and *B*, the DC charging current varies little, less than 1 per cent in a typical case; including the effect of ΔE. It is therefore expected that DPP should be over 100 times more sensitive than convential DC polarography. The rejection of DC charging current improves as Δt decreases.

The transition to the new DC charging current takes place through a transient (Figure 6-8*e*) that is normally quite large but of short duration. Its form follows the behaviour of the double-layer capacitance charged through the resistance *R* of the solution :

$$\Delta I_{chg} = \frac{\Delta E}{R}\exp\left\{-\frac{1}{CAR}\cdot t\right\} \tag{6.7}$$

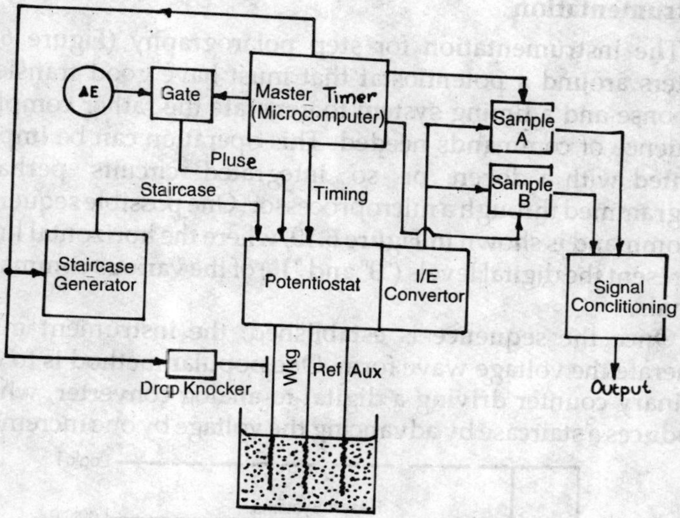

Fig. 6.9. An example of pulse instrumentation. The signal-conditioning module furnishes to the recorder the necessary information concerning ΔI and ΔE.

where C is the capacitance per unit area. The exponential dependence is such that the value decreases rapidly for $t > CAR$. In a typical case, CAR might be about 0.1 ms, so that beyond 1 ms or so the charging transient will have become negligible.

The total current is the sum of those previously discussed and is shown in Figure 6.8 *f*. The dominant contribution to ΔI will be the faradaic transient, with practically no charging current and better than 99 per cent rejection of both the diffusion currents from previous reductions and the charging current due to drop growth. A yet smaller contribution derives from the fact that the potential at B is ΔE volts more negative than at A. This is reflected in both DC faradaic and charging components.

These four residual currents represent the limiting factors in both differential pulse and square-wave polarography. Of the two, square-wave is favoured in terms of sensitivity, since it uses shorter times between A and B. In addition, as shown by Rifkin and Evans, the useful signal increases for reversible systems of Δt is small. Unfortunately, there has always been much less interest on the part of industry in providing square-wave polargraphs than different pulse units.

Instrumentation

The instrumentation for step polarography (Figure 6.9) centers around a potentiostat that must have good transient response and a timing system to generate the rather complex sequence of commands needed. This operation can be implemented with a dozen or so integrated circuits perhaps programmed through a microprocessor. One possible sequence of command is shown in Figure 6.10, where the horizontal lines represent the digital levels ("0" and "1") of the various command signals.

Once the sequence is established, the instrument must generate the voltage wave form. One popular method is to use a binary counter driving a digital-to-analog converter, which produces a staircase by advancing the voltage by one increment

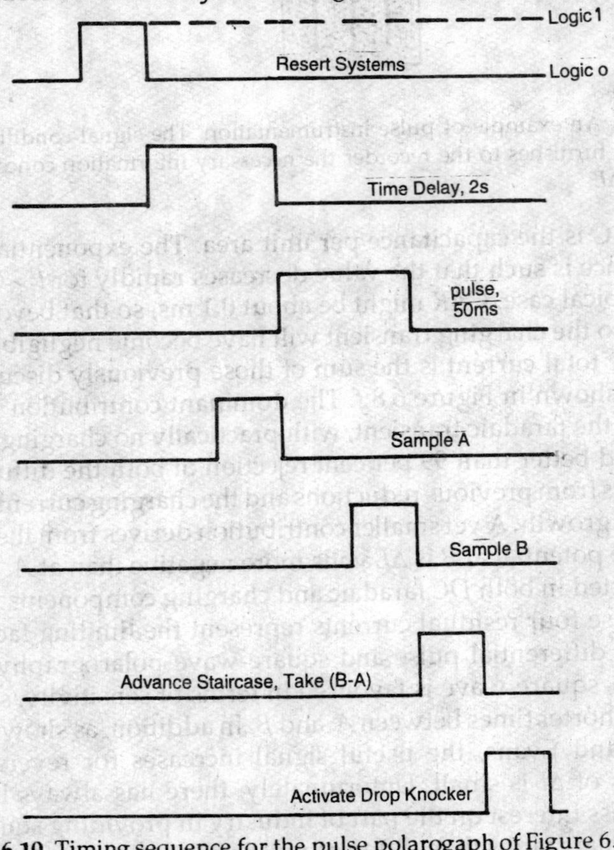

Fig. 6.10. Timing sequence for the pulse polarogaph of Figure 6.9.

every time a count is eceived. Synchronized with each step is a command that allows the voltage pulse, ΔE, to be applied at the proper time and other commands to ensure that the current is read at the correct points, *A* and *B*. These signals are implemented through analog transmission gates.

The current values at times *A* and *B* are fed into two sample-and-hold amplifiers that store the corresponding values, passing them on in turn to a subtraction element to give the desired output, s *I*, to a recorder.

7

Voltammetry : AC Polarography

It was remarked previously that voltammetry can be considered to be largely the study of the equivalent circuit of the electrode-solution interface. Since this interface has significant capacitive properties, more information can be obtained from the electrode *impedance* than from its resistance alone. To make such a study, we can apply a small alternating signal to the cell along with the usual DC ramp or staircase and examine its effects on the cell current. There are a number of closely related techniques that make use of this approach; they are special cases of the field of AC polarography.

The signal-to-noise (S/N) relations are favourable because the information about the chemical system is carried in a narrow band of frequencies only, and extraneous signals at other frequencies can be effectively eliminated. This is an example of AC modulation, much used in various fields to increase the S/N ratio.

We shall see that AC polarography permits an increase in sensitivity compared to the conventional DC method, so that the detection limits in favourable cases can be reduced to about 10^{-8} *M*. An interesting feature is that the sensitivity decreases drastically as the redox system becomes less reversible. This may be advantageous if a reversible system is partially masked by one that is irreversible.

The Basic Experiment

The electrodes and cell configuration of DC polarography are usable in the AC mode without modification. Let us apply an excitation to the cell combining the DC staircase with a small AC voltage (say 10 mV at 150 Hz). The electronic system will extract the AC component of the cell current and display this response as a function of the DC voltage.

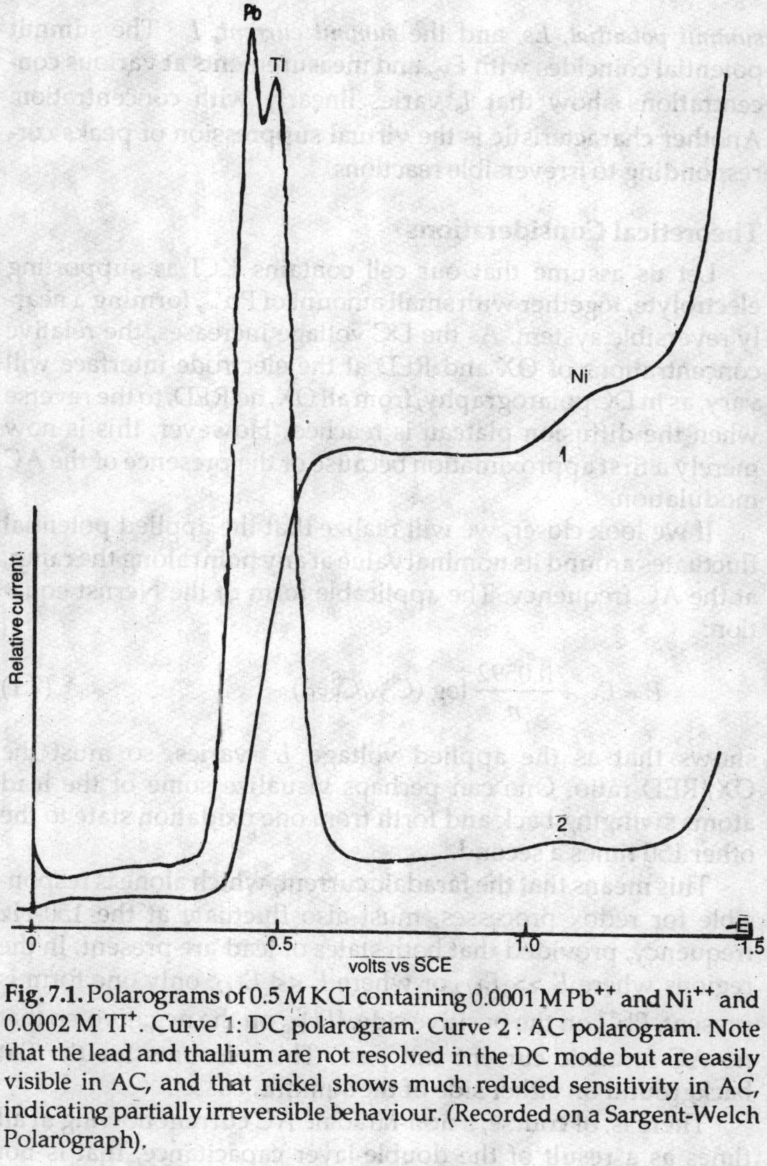

Fig. 7.1. Polarograms of 0.5 *M* KCl containing 0.0001 M Pb^{++} and Ni^{++} and 0.0002 M Tl^{+}. Curve 1: DC polarogram. Curve 2 : AC polarogram. Note that the lead and thallium are not resolved in the DC mode but are easily visible in AC, and that nickel shows much reduced sensitivity in AC, indicating partially irreversible behaviour. (Recorded on a Sargent-Welch Polarograph).

One effect of the AC modulation is to convert the polarograin from the step-like waves, seen in the DC mode, to a derivative curve, similar to those encountered in DPP. Figure 7.1 shows the output signal of such an experiment, compared with a conventional polarogram. The maximum of each peak is called a *summit*, and the corresponding co-ordinates are the *summit potential, Es,* and the *summit current,* I_s. The summit potential coincides with $E_{1/2}$, and measurements at various concentrations show that I_s varies linearly with concentration. Another characteristic is the virtual suppression of peaks corresponding to irreversible reactions.

Theoretical Considerations

Let us assume that our cell contains KCI as supporting electrolyte, together with small amount of Pb^{++}, forming a nearly reversible system. As the DC voltage increases, the relative concentrations of OX and RED at the electrode interface will vary, as in DC polarography, from all OX, no RED, to the reverse when the diffusion plateau is reached. However, this is now merely a first approximation because of the presence of the AC modulation.

If we look closer, we will realize that the applied potential fluctuates around its nominal value at any point along the ramp, at the AC frequency. The applicable form of the Nernst equation:

$$E = E_{1/2} + \frac{0.0592}{n} \log \left(C^S_{OX}/C^S_{RED}\right) \quad (7.1)$$

shows that as the applied voltage E varies, so must the OX/RED ratio. One can perhaps visualize some of the lead atoms swinging back and forth from one oxidation state to the other 150 times a second.

This means that the faradaic current, which alone is responsible for redox processes, must also fluctuate at the 150-Hz frequency, provided that both states of lead are present. In the regions where $E >> E_{1/2}$ or where $E << E_{1/2}$, only one form is present, Pb^{++}_{aq} on the positive side, {Pb^o_{Hg} on the negative, so that no AC faradaic current can pass. This accounts for the flat background on either side of the summit.

There is, of course, a non-faradaic AC current flowing at all times as a result of the double-layer capacitance, that is not affected by the faradaic process. The net alternating current is the sum of the faradaic and non-faradaic components, and this will go through a maximum at the half-wave potential, where the concentrations of OX and RED are equal. The resulting AC polarogram is that already seen in Figure 7.1. Just as in DC, a

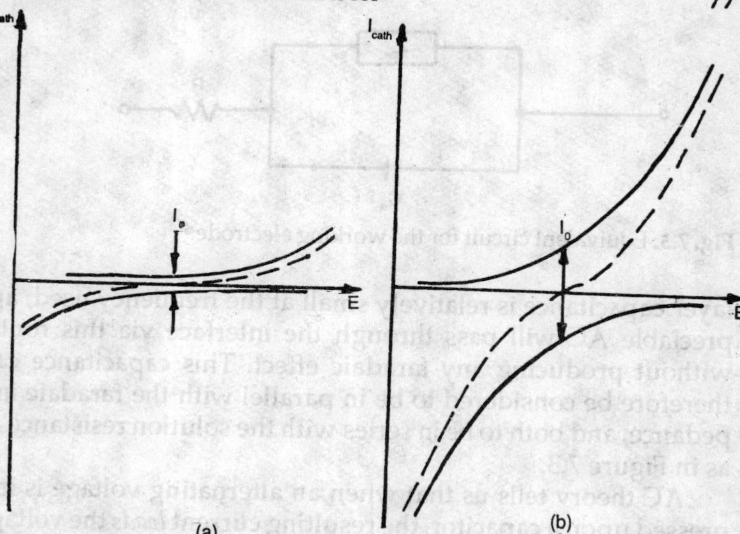

Fig. 7.2. Exchange currents for (a) an irreversible, and (b) a reversible system. The slope of the total current curve (dashed lines) is the faradaic admittance, the reciprocal of the faradiac impedance.

sampling circuit is usually employed to give a step-wise approximation to the envelope.

If, rather than lead, we select nickle to study, the results will be quite different. Since the system is somewhat irreversible, a considerable negative overpotential must be applied to force the reduction to take place, and a similar overpotential of opposite sign is necessary to reoxidize the neutral atoms. This is tantamount to saying that the exchange currents are small (Figure 7.2), whereas for the reversible system they are much greater. Another way of looking at this is to say that in *a* of the figure, the faradaic impedence, Z_{far}, is high, while in *b* it is low. Hence, very little alternating current, Iac $= E_{ac}/Zf_{ar}$, flows in the irreversible case, contrasted with the much greater current if the system is reversible. Experimentally, the response is reduced to a small bump, as seen for nickel in Figure 7.1, or may disappear entirely with a completely irreversible reaction. This effect provides one method of establishing whether or not a system is reversible.

To understand the phenomena taking place in AC polarography, it is essential to grasp the impedance and phase relations at the interface, and this is most readily done by means of an equivalent electrical circuit. Since the impedance of the double-

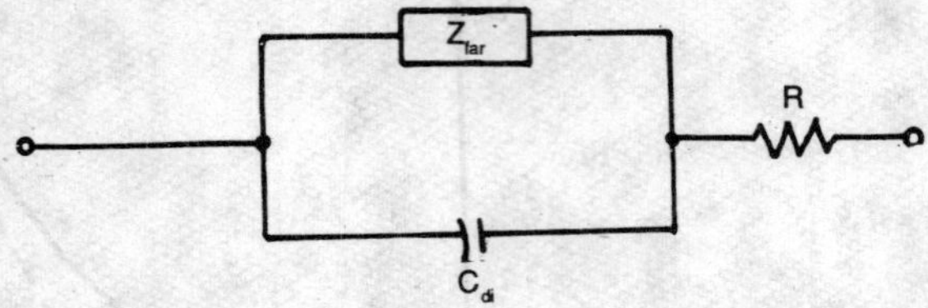

Fig. 7.3. Equivalent circuit for the working electrode.

layer capacitance is relatively small at the frequency used, appreciable AC will pass through the interface via this route, without producing any faradaic effect. This capacitance can therefore be considered to be in parallel with the faradaic impedance, and both to be in series with the solution resistance *R*, as in Figure 7.3.

AC theory tells us that when an alternating voltage is impressed upon a capacitor, the resulting current *leads* the voltage by a phase angle of 90° (Figure 7.4), whereas for a pure resistance the current and voltage are in phase, (0°*c*). In contrast, the faradaic impedance can be shown to produce an intermediate phase shift of 45° (for reversible systems). This circumstance provides a means for distinguishing between faradaic and non-faradaic currents flowing through the equivalent circuit of Figure 7.4.

These phase relations are depicated vectorially in Figure 7.5*a*. The arrow pointing upward represents the applied AC potential and is taken as the reference vector. In the absence of

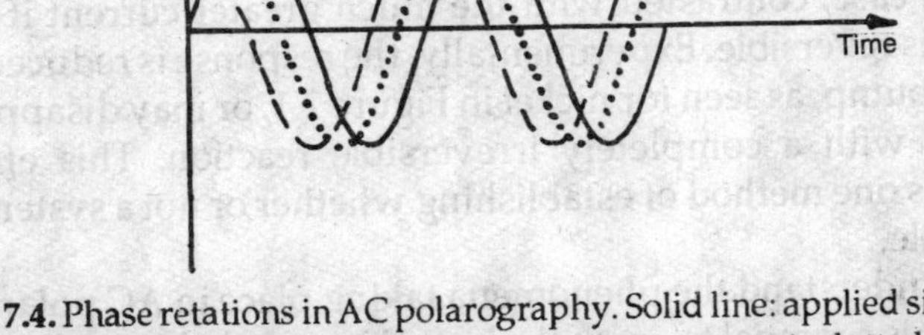

Fig. 7.4. Phase retations in AC polarography. Solid line: applied sine-wave voltage; dashed line: charging current (a cosine wave); dotted line: the faradaic current. The amplitudes are shown as equal for ease of comparison.

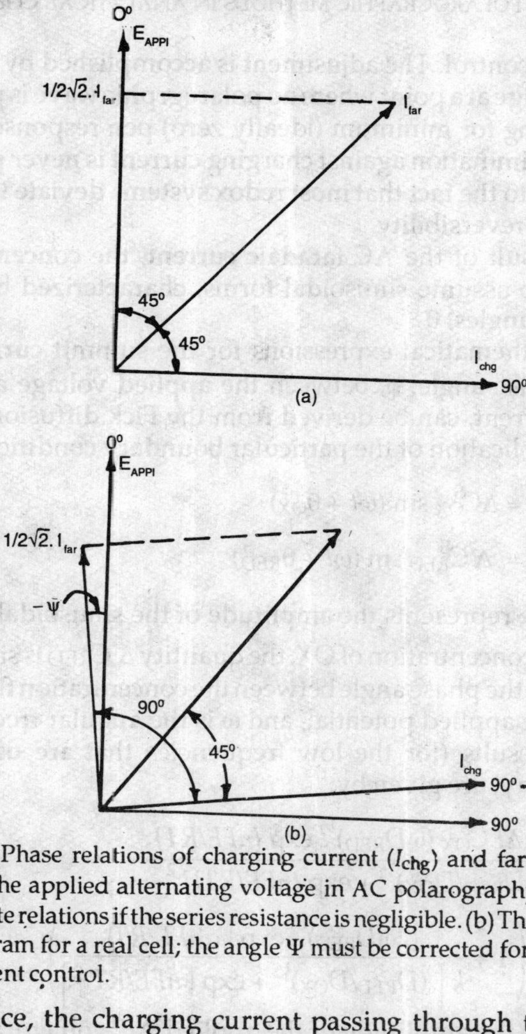

Fig. 7.5. Phase relations of charging current (I_{chg}) and faradaic current (I_{far}) to the applied alternating voltage in AC polarography. (a) The approximate relations if the series resistance is negligible. (b) The corresponding diagram for a real cell; the angle Ψ must be corrected for by the phase adjustment control.

resistance, the charging current passing through the double-layer capacitance is represented by a horizontal vector, and the faradaic current by one at 45°. Hence, an instrument so designed that it responds only to current that is in phase with the applied voltage will, in principle, eliminate the charging current and still give an output that represents 70 per cent ($\sqrt{2}/2$) of the maximum faradaic current.

In the presence of significant resistance, the phase shifts due to both faradaic and capacitive currents are altered by approximately the same amount, Ψ, as shown in Figure 7.5*b*. The discrepancy angle, Ψ, can be allowed for by adjustment of a phase-shift control. The adjustment is accomplished by setting the DC voltage at a point where no polarogrphic wave is present and adjusting for minimum (ideally zero) pen response. Even so, the discrimination against charging current is never perfect, due largely to the fact that most redox systems deviate to some extent from reversibility.

As a result of the AC faradaic current, the concentration profiles also assume sinusoidal forms, characterized by their own phase angles, θ

The mathematical expressions for the summit current, I_s, and the phase angle, ϕ, between the applied voltage and the faradaic current, can be derived from the Fick diffusion equations by application of the particular boundary conditions:

$$C^S_{OX} = \Delta C^S_{OX} \sin(\omega t + \theta_{OX}) \tag{7-2}$$

$$C^S_{RED} = \Delta C^S_{RED} \sin(\omega t + \theta_{RED})$$

where ΔC^S_{OX} represents the amplitude of the sinusoidal fluctations in the concentration of OX, the quantity ΔC^S_{RED} is similarly defined, θ is the phase angle between the concentration fluctutation and the applied potential, and ω is the angular frequency. The final results (for the low frequencies that are useful in polarography) are given by:

$$I_{AC} = \frac{n^2F^2 A\Delta E C^S_{OX} (\omega D_{RED})^{1/2} \exp\{nFE/RT\}}{RT\,[(D_{RED}/D_{OX})^{1/2} + \exp\{nFE/RT\}]^2} \tag{7.3}$$

$$\phi = \cot^{-1}\left(1 + \frac{(2\omega D_{RED})^{1/2} \exp\{\alpha nFE/RT\}}{k\left[(D_{RED}/D_{OX})^{1/2} + \exp\{nFE/RT\}\right]}\right) \tag{7-4}$$

where ΔE is the amplitude of the applied AC voltage, and k is the applicable heterogeneous rate constant.

Certain conclusions can be drawn from these equations. Note that the (faradaic) current in Eq. (7.3) is directly proportional to ΔE as well as to C_{OX}. Also, the current increases with the square root of the frequency (whereas the charging current increases as the first power of the frequency). This would imply that the sensitivity of the method is greater the larger the amplitude and frequency of the AC modulation. However, the derivation of the equations was predicated on both being small enough so that certain approximations could be made. In practice, ΔE is usually between 1 and 30 mV, and the frequency below about 300 Hz. (the power line frequency can be used satisfactorily).

In Eq (7.4), k is normally greater than $(2\omega D_{red})^{1/2}$, so that the second term approaches zero and $\phi \rightarrow \cot^{-1}(1) = 45°$.

Instrumentation

The basic structure of an AC polarograph is shown in Figure 7.6. The AC from the oscillator is combined with the DC ramp or staircase and impressed on the potentiostate, amplifer 1. The cell current, containing both AC and DC components, is converted by amplifier 2 to a corresponding voltage. The combination of R and C acts as a high-pass filter, permitting the AC portion of the signal to pass, but supporessing the DC component. The AC signal is then rectified to give a proportional DC, and recorded.

In order to take advantage of the phase relations demonstrated in Figure 7.5, a phase-sensitive detector must be included, as shown in Figure 7.7. This is a device that compares the AC signal with a reference at the same frequency obtained directly from the oscillator, and acts as a gate to transmit to the recorder only that component of the signal that bears a specific phase relation. In order to compensate for the phase changes introduced into the signal by various electronic components and by the cell itself, an adjustable phase-shifting device must be included in the reference connection.

The phase-sensitive detector also acts as a filter, discriminating against frequencies other than that of the oscillator, including harmonics. Since electrical noise occurs at all frequencies, most of it is eliminated by this filtering action. In this context, the double-layer charging current can be considered to be noise.

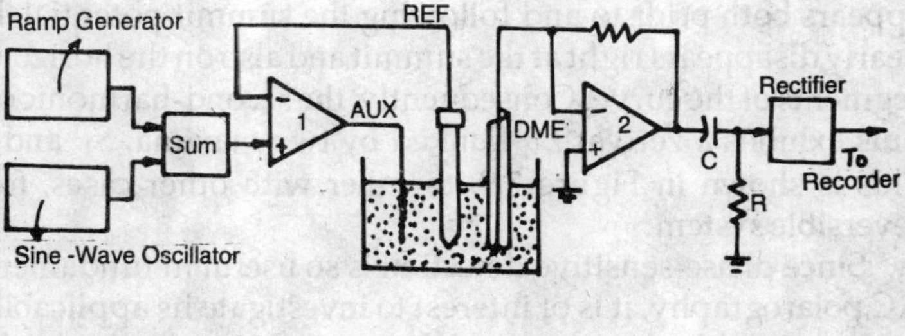

Fig. 7.6. A basic AC polarography.

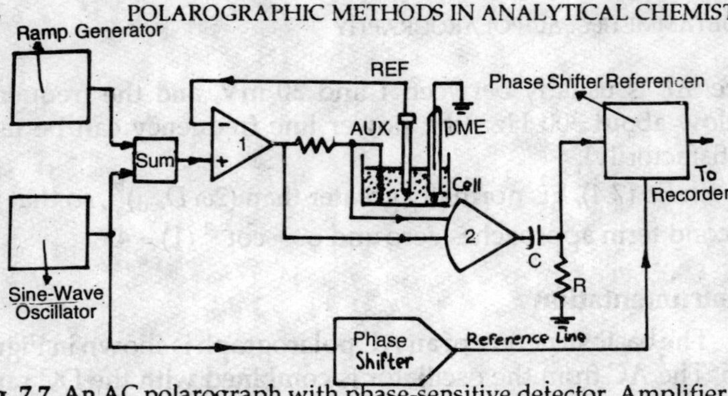

Fig. 7.7. An AC polarograph with phase-sensitive detector. Amplifier # 2 is shown as an instrumentation or differential amplifier, measuring the potential drop across the resistor in the auxiliary electrode line.

Second-Harmonic AC Polarography

When a pure sine-wave voltage is impressed upon an electrical circuit component, the current that passes may or may not also be a sine wave. If the component is linear, then the response is still a sine wave, wherease, for non-linear components, it is subject to *harmonic distortion*, which consists in the generation of multiples of the original frequency. Figure 7.8 shows how distortion can originate in the polarographic cell.

The faradaic impedance at the DME constitutes a non-linear circuit element so that the alternating portion of the faradaic current contains an appreciable percentage at the second harmonic frequency — twice the frequency of the applied voltage. (Contributions of higher harmonics can be neglected). On the other hand, the double-layer capacitance is a nearly linear element, and thus does not contribute appreciable to the harmonic content. This means that a polarograph that responds to the second harmonic, rather than to the entire AC current, will discriminate strongly against the charging current.

The non-linearity that gives rise to harmonic distortion appears both prior to and following the summit potential, but nearly disappears right at the summit and also on the horizontal segments of the curve. Consequently, the second-harmonic outputs exhibits a zero at $E_{1/2}$ flanked by two maxima, S_1 and S_2. This is shown in Figure 7.9, together with other cases, for a reversible system.

Since phase-sensitive detection is so useful in fundamental AC polarography, it is of interest to investigate its applicability to the second-harmonic case. We cannot here take the applied

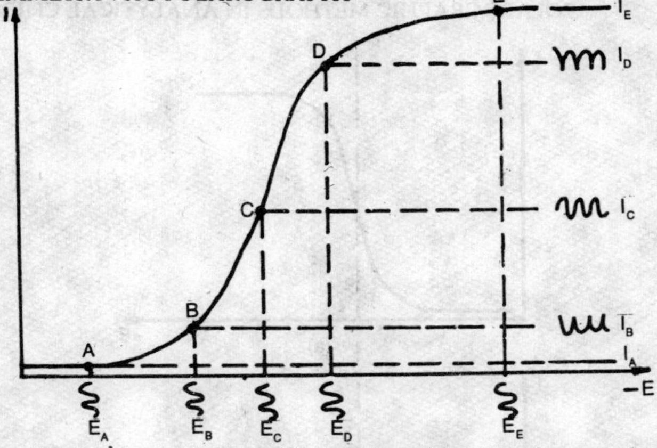

Fig. 7.8. The source of harmonic distortion in AC polarography. Small sinusoidal voltages are shown at five locations. At E_C (same as $E_{1/2}$, the polarogram is nearly linear, and no distortion results. At E_B and E_D, where the curvature is marked, the current shows considerable distortion. At E_A and E_E, the curve is so nearly horizontal that no sinusoidal current variation is seen at all.

alternating voltage as a reference vector because phase differences can only be defined as relation variables of the *same* frequency. Hence, the best we can do is to look at the *relative* phases of the second-harmonic current in the two branches of the curve of Figure 7.9 *c*.

In order to instrument second-harmonic polarography, the basic circuit of Figure 7.7 can be used, with the sole addition of a component to go between the oscillator and the phase shifter, to double the frequency in the reference line. This is shown in Figure 7.10. The result of an experiment with this modified instrument gives a curve such as that in Figure 7.9 *d*. This shows one of the branches as inverted indicating a phase difference of 180° between the branches, in agreement with theoretical predictions. An example is presented in Figure 7.11.

The mathematical theory of second-harmonic AC polarography is even more complex than that of the fundamental case. It leads to equations that are quite intractable unless solved by computer. The final equation shows that the current is indirect proportion to the bulk concentration of the reducible species and to the square of the applied AC voltage amplitude. It also increases as the frequency is increased, but by a non-linear function. Theory correctly predicts that, for a reversible process, the phase-detected current will be of one sign for potentials

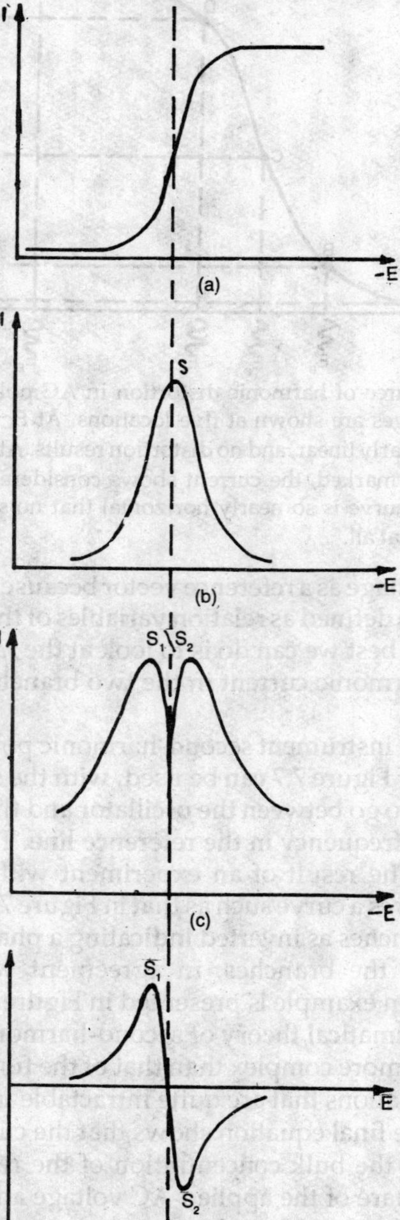

Fig. 7.9. Comparison of sampled polarograms; (*a*) DC, (*b*) simple AC, (*c*) second-harmonic AC, (*d*) second-harmonic AC with phase-sensitive detection.

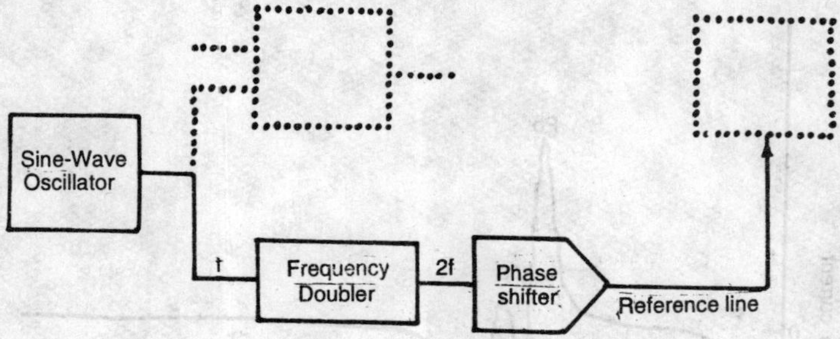

Fig. 7.10. Modification to the circuit of Figure 7.7 to allow for phase-sensitive detection at the second-harmonic frequency.

more positive than the half-wave potential, and the opposite sign for more negative potentials, passing through zero at $E_{1/2}$. The two summits are separated by 68.4/n mV.

The most convenient measure of concentration is the vertical distance between the two summits, since this automatically corrects for any error in placement of the background or residual current level.

The dependence of the output on the square of the applied AC amplitude sugests the benefit of using as large a voltage as possible, but in practice it must be limited to about 10 mV in order to avoid degradation of resolution. The frequency can be varied over a wider range than in the fundamental AC case. For example, the reduction of Cd^{++} to Cd_{Hg} in a neutral medium shows an increase in summit current from about 0.9 μA at 23 Hz to 3.2 μA at 2220 Hz. On the other hand, the same system shows no change in the voltage separation of the two summits. It must be emphasized that the preceding discussion applies only to systems that show reversible characteristics when measured at DC. Others will give complicated deviations from the basic equations [2, 6], and the suitability of the method for specific cases must be determined experimentally.

Methods Related to AC Polarography

As we have seen, an alternating voltage impressed on a non-linear circuit element gives rise to harmonics of the applied frequency. The current can be expressed by a Taylor's series. If the excitation is given by $E_{AC} = E_0 + A \cos \omega t$, then the AC response is $I_{AC} = E_{AC}/Z_{far}$. Expansion then gives:

Fig. 7.11. Second-harmonic, phase-detected AC polarogram. The solution is the same used for Figure 6.1. Note that the nickel does not show up at all. (Recorded on a Sargent-Welch Polarograph).

$$I_{AC} = b_1 \cos (\omega t + 45^o) + b_2 \cos^2 (\omega \tau + 45^o)$$
$$+ b_3 \cos^3 (\omega t + 45^{o)} + \cdots \quad (7.5)$$

where the *b*s include all constant coefficients. The angle 45° is the phase shift due to the faradaic impedance. The series rapidly converges, which means that $b_1 > b_2 > b_3$...., and for the present case, the third and higher harmonic terms can be neglected. The second term can be restated in a more useful form by making use of the trigonometric identity.

$$\cos^2 x = \frac{1}{2} + \frac{1}{2} \cos 2x \quad (7.6)$$

so we can write :

$$I_{AC} = b_1 (\omega t + 45^o) + \frac{1}{2} b_2 + \frac{1}{2} b_2 \cos (2 \omega t + 90^o) \quad (7.7)$$

The first term refers to the fundamental AC polarographic signal. The last term, in cos 2 ωt, describes the second harmonic signal. This leaves the invariant term $\frac{1}{2} b_2$, which must correspond to a DC component of the current.

This can be visualized by reference to Figure 7.12. Since this DC component and the term in $2\omega t$ arise from the cosine-squared term of the expansion, they should be expected to carry the same information about the electrochemical system. This is a form of rectification, and its measurement is called *faradaic rectification polarography*.

Another method that utilizes the non-linear nature of the electrode process is known as *intermodulation polarography*. This requires the simultaneous application of two different frequencies, f_1 and f_2. The action of the non-linear circuit is to generate the sum and difference frequencies, $f_1 + f_2$ and $f_1 - f_2$. For example, if two oscillators connected to the input of the potentiostat are set at 23 Hz and 300-Hz, respectively, then the cell current will contain components at 277 and 323 Hz in addition to the unchanged 23- and 300 Hz frequencies. Measurement with the aid of a suitably tuned amplifier and detector provides information similar to that of second-harmonic AC polarography.

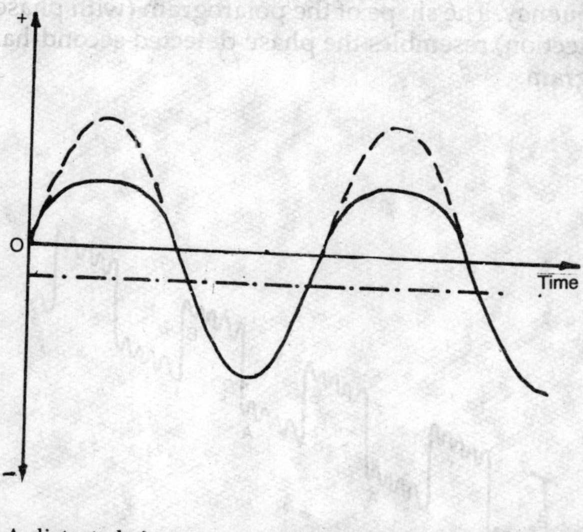

Fig. 7.12. A distorted sine wave resulting from a decreasing slope of the current-voltage curve. The dashed curve is the original sine wave, the solid curve shows the distortion effects. The dash-not line represents the DC current due to faradaic rectification. Additional components of higher frequencies are also generated.

For analytical purposes, these methods have little advantage.

AC-Pulse Polarography

It is possible to combine the principles of AC polarography with those of pulse techniques in such a way as to retain the beneficial features of both. This has been reported for the "normal" pulse case, but to date has been found most useful with the differential pulse method.

The potential applied to the electrode as a function of time is shown in Figure 7.13. This can be regarded as AC polarography wherein the DC ramp is replaced by the pulsed ramp as used in DPP. The requirements are simply that there be many cycles of AC modulation uring each pulse, and that the AC amplitude, ΔE, be less than the pulse height, E_p.

The measurement procedure, just as in DPP, is to sample the current (AC in this case) just *before* the pulse (I_a) and again near the end of the pulse (I_b). The theory then shows that the difference $I_b - I_a$ is proportional to the bulk concentration of the active species, to the amplitude ΔE, and to the square root of the frequency. The shape of the polarogram (with phase-sensitive detection) resembles the phase-detected second-harmonic polarogram.

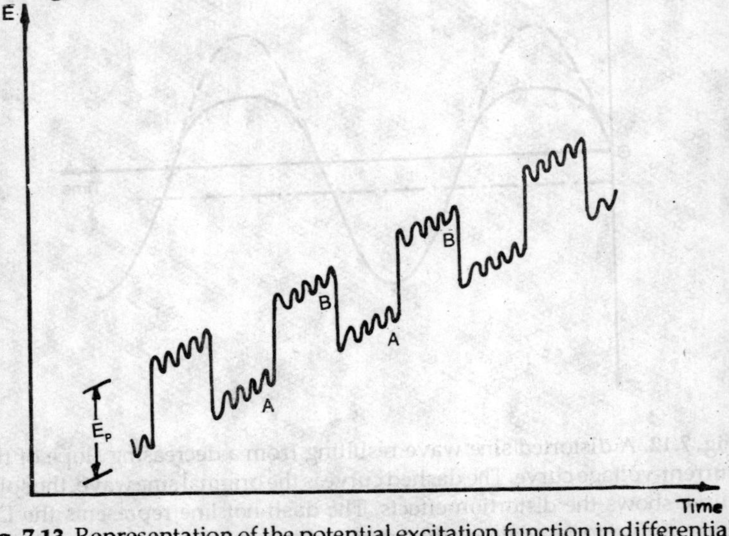

Fig. 7.13. Representation of the potential excitation function in differential pulse AC polarography. Measurements are taken at points *A* and *B*.

The charging current produced by the pulse almost completely dies out before the current is sampled, as in DPP. The effect of the small residuum is merely to offset the background level slightly, and even this effect is eliminated by making quantitative measurements vertically between the two summits.

This makes another sensitive and useful method for the determination of low concentrations of species with reversible or quasi-reversible electrode processes. The limit of detection for a process such as the reduction of Cd^{++} in KNO_3 is reported to be about 5×10^{-8} *M*. It is not applicable to irreversible systems.

8

Analytical Applications of Polarography

General

The polarographic wave for a solution containing a single reducible species may be analyzed to determine the half wave potential and with a consideration of the supporting electrolyte involved, the species may be identified. Mixture of reducible species may also be analyzed and various species indentified in the same way, provided that the half wave potentials are separated by about 0.2 to 0.3 V, even when only one of the adjacent half-waves represents a reaction with n=1 and by 0.1 to 0.2 V where the adjacent half-waver represent reaction each with n=2. Although the qualitative aspect of polarography should not be over looked. By for the most important analytical application of technique in that having to do with the quantitative determination of a species or the components of a mixture of suitable species. The Iikovic equation is the basis of quantitative aspects, since all other components of the expression being accounted for, the value of I_d will be directly proportional to the concentration of reduciable (cathodic) or oxidisable (anodic) species. In order to evaluate concentration by polarographic technique, it is necessary to determine the plateau or diffusion current as the time average diffusion current $\bar{i}d$

Various applications of polarography have been studied under various heads such as inorganic polarography, organic polarography, amperometric titrations, valtametry etc.

In this chapter we will denote our attention to organic polarography.

ORGANIC POLAROGRAPHY

General

Most organic electrode reactions are ineversible and majority of these take place in stepwise fashion. In very many instances these reactions are pH dependent. Where H^+ is consumed or produced by the electrode reaction, the solution must be adequately buffered to prevent undesirable $[H^+]$ values at the interface as the reaction proceeds. In most cases solvents other than water must be used and this often requires that corrections be made because of high solution resistance and associated high i_R drop through the solution. Because of these characteristics the polarographic data obtained is sometime difficult to interpret, Never the less, many organic compounds can be treated polarographically with respect to qualitative identification, quantitative estimation and structural elucidation several functional groups react at DME and compounds containing these groups may often be treated polarographically. Such groups include organic halogen compounds, cabonyl compounds, including ketones, sugars, aldehydes, quinones, and hydroquinones. Same carboxylic acids and dicarboxylic acids, excepting simple aromatic and aliphatic monocarboxylic acids; same keto and aldehydo acids, nitro, nitroso, azo and amine oxide group compounds; oxygen-oxygen, sulphur-sulphur and carbon-sulphur bond substances.

A carbon-carbon double bond in conjugation with another double bond, an aromatic ring system or another unsaturated group is reducible at DME. For example, the reduction of stilbene has been proposed to occur according to the following sequence of reactions:

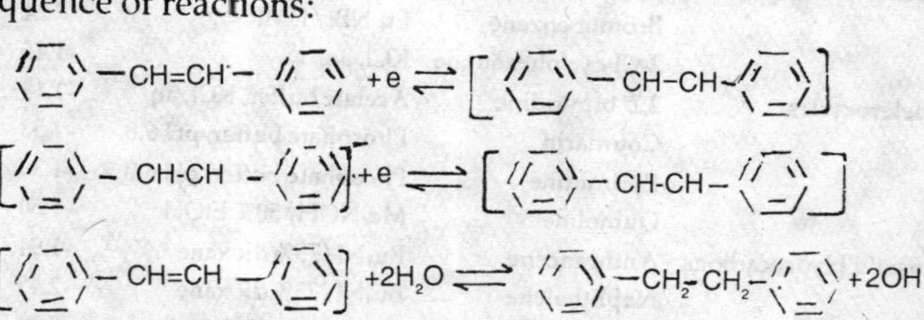

Organic halogen compounds undergo reduction with the consequent replacement of the halogen with hydrogen, as exemplified by the reaction of ethyl iodide.

$$CH_3CH_2I + H^+ 2e \rightarrow CH_3CH_3 + I^-$$

carbonyl compounds, including aldehydes, ketones, ketosteroids, and sugars are irreversibly reduced at a mercury cathode, where as the reduction of quinons and the oxidation of hydroquinones proceed reversibly.

A classic example of the latter set of reaction is the inter convertibility of *p*-benzoquinone and *p*-benzohydroquinone

– Dicarboxylic acids, keto acids, aldehyde acids and halogenated acids can be reduced.

A representative list is given in Table 8.1. To be reducible under polarographic conditions, a compound must possess one or more polar or unsaturated linkage. The half-reactions often require H^+ ions for completion. As an example, the two-step reduction of α -aminoacetophenone follows the mechanism.

TABLE 8.1
Some Organic Compounds Reducible at the DME[a]

Class	*Example*	*Medium*	$E_{1/2}$	*n*
Carboxylic Acids	Acetic acid	$Et_4 NClO_4$/MeCN	-2.3	
	Fumaric acid	NH_3 buffer/ 10% EtOH	-1.57	
	Maleic acid	NH_3 buffer/10% EtOH	-1.35	
	Oxalic acid	Et_4NClO_4/MeCN	-1.6	
Esters	Ethyl acrylate	Me_4 NI/30% EtOH	-1.82	
	Diphenyl phthalate	Me_4 NCl/75% EtOH	-1.65	
Carbonyl Compounds	Acetaldehyde	LiOH, aq.	-1.93	
	Acetone	Bu_4 NCl, BuNOH/80% EtOH	-2.53	2
	Me_2- glyoxime	HCl/15% EtOH	-0.81	8
Halogen Compounds	Allyl bromide	Li_3 Cit/50 dioxane	-1.18	2
	Bromobenzene	Et_4 NBr/DMF	-2.24	2
	2-Cl-cyclohexanone	KCl, aq.	-1.45	
Heterocycles	2,2′ bipyridine	Acetate buffer, KCl, aq	-1.14	1
	Coumarin	Phosphate buffer, pH 6.8	-1,53	2
	Pyrimidine	Phosphate buffer, pH 6.8	-1.30	2
	Quinoline	Me_4NOH/50% EtOH	-1.50	2
Insat'd Hydrocarbons	Anthrancene	Bu_4NI/75% dioxane	-1.94	2
	Naphthalene	Bu_4NI/75% dioxane	-2.49	2
	Phenanthrene	Bu_4NI/75% dioxane	-2.44	2
	Styrene	Bu_4NI/75% dioxane	-2.34	2
Nitro Compounds and related	Azobenzene	NaCl, KCl/30% MeOH	-0.81	2
	Ethyl nitrate	LiCl/10% EtOH	-0.82	2
	Nitrobenzene	Acetate buffer/60% EtOH	-0.43	
	Trinitroglycerol	Me_4 NCl/75%/ EtOH	-0.70	
Sulfur Compounds	Cystine	NH_3 buffer /pH 9.2	-1.3	
	Diphenyl-sulfoxide	Me_4 NBr/50% EtOH	-2.07	2
	Thiourea	H_2SO_4 aq.	(+0.04)	
	(anodic wave; forms ppt with Hg)			

Note: Reference electrodes must be selected with care, to be compatible with organic solvents.

$$\phi - CO - CH_2 - NR_3 \xrightarrow{H^+, 2e} \phi - CO - CH_3 \xrightarrow{2H^+, 2e^-} \phi - CHOH - CH_3$$

This was observed in a pH 7 phosphate buffer, so the required hydrogen ion was merely that available from the dissociation of solvent water.

Organic applications have been reviewed extensively by Zuman and by Elving.

Solubility is, of course, a source of difficulty with organic compounds. Many materials usually regarded as insoluble in water are, nevertheless sufficiently soluble to be measurable polarographically. Others are easily accessible in mixed solvents, such as 50 per cent aqueous alcohol or dioxane, but many require anhydrous solvents such as aceonitrile or dimethyl formamide. In aqueous solvents, a number of the familar salts or acids are sufficiently soluble and ionized to serve as supporting electrolytes. In anhydrous media, tetra-alkylammonium salts are often the only suitable materials for this purpose.

The choice of reference electrodes is another problem with nonaqueous solvents, whether they are truly anhydrous or not. Frequently, a conventional calomel or silver chloride is selected with a liquid junction between aqueous KCI solution and the prevailing solvent. Such an electrode may serve empirically as a reference, but the expected large and unknown junction potential precludes accurate and reproducible data on half-wave potentials.

Various reference electrodes have been described for use in specific solvents. In many, a silver wire immersed in a solution of Ag^+ ions in the organic solvent will suffice. McMaster et al. prepared a stable electrode for use in dimethylsulfoxide, consisting of a saturated solution of zinc perchlorate in contact with a saturated zinc amalgam. Recent work by Peerce and Bard has shown that a platinum electrode coated with a polymer of vinylferrocene serves as a good reference in acetonitrile, and suggests that polymer electrodes of other composition may be similarly useful. Other reference electrodes are described by Ives and Janz and by Butler.

9

Applications of polarography in Inorganic Analysis

Scope of Applicability

Polarographic analysis is applicable to many kinds of substances. The principal requirements are that a substance be soluble in an ionizing solvent and that it be oxidizable or reducible at a potential within the range accessible to the DME, namely about +0.4 to beyond -2 V (in aqueous solution, vs. SCE). The electroactive substance may be an ion of either change or a neutral species.

The majority of polarographic analyses involve reductions, and nearly any metal, except perhaps the most active, can be reduced from its salts or complexes, either to the metal or to the lower oxidation state. Double waves are often seen with species such a chromate, that can be reduced in two steps.

Anodic oxidations that are observable polarographically include metallic species that exist in a lower oxidation state in solution. Occasionally a useful anodic wave can be obtained that involves oxidation of mercury. For example, traces of chloride can be determined by the reaction.

$$2Hg + 2Cl^{-} - 2e^{-} = Hg_2Cl_2$$

Determination of Metal Ions

Through an appropriate choice of supporting electrolyte, pH and other complexing ligands, practically any metal ion can be reduced at the dropping mercury electrode to an amalgam or to a soluble lower oxidation state. In the majority of situations, one obtains polarographic waves and diffusion currents suitable for quantitative determination of these species common divalent cations such as cadnium, cobalt, copper, lead, manganese, nicket, tin and zinc can be determined in many different complexing and non-complexing media. The alkaline earth ions-barium, calcium, magnesium and strontium exhibit well defined polarographic waves at approximately 2.0 V vs SCE in aqueous and alcoholic tetra alkylammonium hydroxide media. Polarographic characteristics of the tripositive states of aluminium, bismuth, uranium, vanadium and yeterbium in a number of supporting electrolytes have been reported.

Determination of Inorganic Anions

Polarographic studies of reduction of chromates, Iodates, molybdates, selenite, tellurite, and varadate as well as the anionic chloride complexes of tungsten (vi), tin (iv) and molybdenum (vi) are included among the list of analytically useful procedures. In the presence of bromide ions, The DME is oxidised to insoluble mercurous bromide.

$$2Hg + Br^- \rightleftarrows Hg_2 Br_2(s) + 2e$$

and a polarographic diffusion current proportional to the bulk concentration of the anion is observed. Other species, including chloride, iodide, sulphate, and thiocyanate, exhibits analogous behaviour. In addition, cyanide and thiosulphate both form soluble complex ions with mercury (II)

$$Hg + 4CN^- \rightleftarrows Hg\,(CN)_4^{=} + 2e$$

$$Hg + 2S_2O_2{}^{--} \rightleftarrows Hg\,(S_2O_3)_2{}^{--} + 2e$$

An anodic polarogram for the oxidation of mercury in a dilute sodium cyanide medium shows a limiting current governed by diffusion of cyanide to the surface of the dropping mercury electrode.

Analysis of Mixtures

Solutions containing two or more electroactive substances may be analyzed polarographically provided certain requirements are satisfied. Clearly, the half-wave potentials for the species of interest must differ significantly if the resulting polarogram is to exhibit a distinct polarographic wave and limiting current region for each component. The minimal difference in the halfwave potentials may also be established.

A mixture of cadmium (II), zinc (II), and cobalt (II) in a IF potassium chloride medium can be resolved and analyzed by means of polarography, since Table 9.1 indicates that the respective half-wave potentials for these elements are –0.60, –1.00, and

1.20 V versus SCE. However, difficulty in the accurate analysis of such a mixture can be encountered if the relative concentrations of the three cations are widely disparated. For example, if a small quantity of zinc is present in a solution containing relatively large amounts of cadmium and cobalt, the polarographic wave for reduction of zinc will appear as a tiny hump on top of the cadmium wave and will be closely followed by another large rise in current because of cobalt (II) reduction. One remedy for this situation involves polarographic measurement of cadmium (II) and subsequent removal of most, if not all, of this ion by contolled-potential separation. Then, zinc may be more accurately determined because the interference from cadmium is eliminated. Furthermore, if cadmium (II) is separated quantitatively and a very low concentration of zinc (II) is present, one may measure the height of the polarographic wave for the cobalt (II) zinc (II)-mixture and calculate the concentration of cobalt by ignoring the small diffusion current due to zinc reduction.

TABLE 9.1. Half-wave Potentials for Common Metal Ions*

Half-Reaction	*Supporting Electrolyte solution*	*E½ Versus SCE, V*
Bi (III) → Bi(Hg)	I *F* HCl	–0.09
Cd (II) → Cd (Hg)	0.1 *F* KCl	–0.60
Co (II) → Co (Hg)	I *F* KCl	–1.20
Cu (II) → Cu (Hg)	0.1 *F* KNO_3	+0.02
Cu (II) → Cu (Hg)	I *F* KCl	+ 0.04
Cu (I) → Cu (I)	I *F* KCl	–0.19
pb (II) → Pb (Hg)	0.1 *F* KNO_3	–0.39
TI (I) → TI (Hg)	0.1 *F* KNO_3	–0.46
Zn (II) → Zn (Hg)	1 F KCI	–1.00

*Data excerpted from L. Meites: Polarographic Techniques, Second Edition, Willey-Interscience, New York, 1965, pp. 615-670. This reference work contains a wealth of information pertaining to the polarographic behaviour of inorganic and organic compounds.

Another procedure for the polarographic analysis of mixtures takes advantage of the fact that half-wave potentials, like standard and formal potentials, for reductions of metal ions change in the presence of complexing agents. As an example, the individual determinations of lead (II) and thallium (I) in a mixture cannot be accomplished if the supporting electrolyte solution is 0.1 *F* potassium nitrate because the half wave potentials are too close together, being –0 and – 0.46 V versus SCE, respectively (see Table 9.1). However, in a I *F* sodium hydroxide medium containing a small amount of maximum suppressor, the reduction of thalium (I) yields a well-developed polarographic wave at –0.48 V versus SCE, whereas reduction of lead (II) occurs at a half-wave potential of –0.76 V versus SCE. Under such conditions, both species may be determined accurately. It should be noted that the half-wave potential for thallium (I) reduction shifts very little, suggesting the absence of strong complex formation in either potassium nitrate or sodium hydroxide media. On the other hand, the large change in the half-wave potential for the lead (II) wave is in agreement with the known fact that the hydrogen plumbite anion, $HPbO_2$, is formed in alkaline solution.

Determination of the Purity of Reagents

Polarography is a very useful technique for indentifying and estimating electroactive impurities specially in electroinactive materials. Thus traces of As, Pb, Fe, Mn etc., can be easily identified and estimated in a sample of alkali or alkaline earth salt or in acids like HNO_3, H_2SO_4 etc. However, any non-reducible impurity present in these regents cannot be detected by polarographic analysis. Thus the presence of $MgCl_2$ or $CaCl_2$ in a sample of NaCl or KCl cannot be detected polarographically. Similarly one can easily detect the presence of an electroactive inorganic or organic component in an electroinactive organic reagent such as EDTA. However, the presence of an amine or carboxylic acid, as an impurity, cannot be detected polarographically since neither amines nor carboxylic acids are reducible at the DME. If the substance whose purity is to be determined is electroactive, then only such impurities present in it, which give waves preceding the wave of the active substance, can be identified and estimated. The percentage purity of this electroactive substance can also be determined, in the absence of any interferring waves, by obtaining a polarogram of a solution containing a known amount of the substance and determining the concentration of the active constituent from its wave height, by one of the standard methods. The error involved in this method of determing the purity would be of the order of 2.3%. Classical methods, like gravimetric and volumetric analysis would normally give a much higher accuracy. The polarographic method, therefore, is very useful for detecting and estimating the presence of electroactive impurities in a reagent, but not very useful for determining its percentage purity.

Inorganic Polarography

The formation of an EDTA complex with a metal ion usually results in a shift of the half-wave potential of the polarographic wave of the metal. This displacement in the half-wave potential has been extensively used in the polarographic determination of various metals. It is possible to separate the waves of two or more metals having similar reduction potentials by forming complexes. The cause of the spread in half-wave potentials for the complexes is due in part to the relative strengths of the bonds between the metals and the complex-former. Pribil and co-workers have used this principle to resolve the waves of copper, (II) lead (II) and thallium (I):

A number of polarographic titration methods are described in the following pages:

Determination of Barium and Sulphate

Tockstein and Serak, have descibred an interesting polarographic method for the determination of barium and sulphate. This determination is based on the properties of a mixture of free silver ions with an excess of the silver complex with EDTA in ammoniacal solution. This mixture yields two polarographic waves, the more positive corresponding to the concentration of free silver ions, and the more negative being proportional to the silver -EDTA complex concentration.

When barium sulphate in dissolved is an ammoniacal solution of the silver complex containing a small amount of free silver ions, the more positive wave increases. The increase is cause by the formation of the barium complex with EDTA, which is stronger than the silver complex, so that an equivalent amount of free silver ions is liberaed. Therefore, the change in height of the first wave is proportional to the amount of barium sulphate dissolved.

Procedure

Precipitate barium or sulphate from a solution containing the equivalent of 7 mg of sulphate ion or less per 10 ml. Wash the precipitate several times with hot water, centrifuge, and treat with 10 ml of a solution that is 0.01 *N* is silver -EDTA complex, I *N* in ammonium hydroxide, and which contains 3-5 per cent free silver ions. After the barium sulphate has dissolved in the solution, mix the latter with 10 ml of 1 *N* ammonium hydroxide, 2.5 ml 0.001 *M* fuchsin, and 1 ml of 0.5 per cent gelatin, and dilute with water to 50 ml. Measure the solution polarographically in an atmosphere of nitrogen, using a mecrurous sulphate reference electrode.

This method has been applied to the analysis of ores and India rubber.

Determination of Bismuth

Zabransky has used disodium EDTA in a polarographic method for the determination of small amounts of bismuth in copper. This method consists in the polarographic determination of bismuth after separating from copper by extraction with dithizone.

Procedure

Dissolve 5 g of copper shavings containing at least 0.0005 per cent bismuth in 50 ml of concentrated nitric acid, and evaporate the solution with 8 ml of concentrated sulphuric acid until white fumes of sulphuriroxide are evolved. Cool, add 50 ml of water, and 1 ml of 50 per cent citric acid. Then add concentrated ammonium hydroxide until the precipitate that forms redissolves. Mix the solution with 50 ml of 50 per cent potassium cyanide, dilute to 200 ml, and extract with a 0.5 per cent solution of dithizone in chloroform. Extract 6 times with 3 ml portions. Evaporate the combined extracts to dryness, heat the residue with 0.5 ml of concentrated sulphuric acid, and 30 per cent hydrogen peroxide to destory organic matter. Expel the excess peroxide, cool, and dilute with 5 ml of 2 M sodium acetate containing 0.05 *M* disodium EDTA and Polarograph. Under the given conditions, the half-wave value of bismuth is approximately – 0.6 volt, and that of lead, which is also extracted by dithiozone, is – 1.1 volts.

Funcherre and Souchay have described a polarographic method for the determination of traces of bismuth in alloys containing large quantities of tin, lead, antimony.

Pribil and Matyska have described a method for the amperometric titration of bismuth in lead.

Procedure

Dissolve the lead sample in nitric acid, and to the strongly acidic solution add sufficient tartrate or citrate that, on dilution, the concentration of the anion is 0.3-0.5 *M*. Dilute to 100ml and adjust the acidity with ammonium hydroxide to pH of 2. Titrate an aliquot part of the solution with 0.02 *M* EDTA at – 0.18 to –0.22 volt vs saturated calomel electrode (SCE) to a constant galvanometer reading. The lead nitrate serves as the supporting electrolyte.

By this method as little as 2 mg of bismuth in 50 mg of lead, corresponding to 0.004 per cent bismuth, can be determined. Tin interferes.

Ishibashi and co-workers have studied the use of an EDTA potassium cyanide solution as supporting electrolyte in polarographic analysis. This method has proved effective for the analysis of bismuth in the presence of a large amount of copper.

Determination of Cadmium

Faucherre and Sounchay have described a polarographic method for the determination of traces of cadmium in alloys containing large quantities of tin, lead and antimony.

Polarographic and Amperometric Determination of Calcium

Pribil and Vicenova have shown polarographically that zinc is quantitatively displaced from its complex with EDTA by calcium ion in 4 *N* ammonium hydroxide solution. This reaction has been made the basis for a very rapid method for the indirect polarographic determination of calcium, even in the presence of small amounts of barium, which must be precipitated as the sulphate. Magnesium interferes.

Procedure

To 2-5 ml of 0.1 *M* zinc-EDTA complex, add 1 ml of sample containing the calcium ion. Dilute to 10 ml with ammonium hydroxide so that the final concentration of the base is 4 *N*. Add a little gelatin, and measure the solution polarographically in an atmosphere of nitrogen. A bank is necessary for the determination. To determine calcium in a solution containing strontium and barium, the latter must be removed by precipitation.

Procedure

To 10 ml approximately 0.005 *M* sample solution, add 5 ml of 0.01 *M* zinc-EDTA complex, Make the solution slightly basic with ammonium hydroxide, dilute to 40 ml, and heat to boiling. Add, at the bolling point, sufficient 0.2 *M* ammonium sulphate to precipitate all strontium more and barium, Filter the precipitate, make the filtrate basic with ammonium hydroxide to form 100 ml of 4-5 *N* ammonium hydroxide solution, and measure polarographically.

Calcium has been determined polarographically in urine and blood. Interference by magnesium is prevented by precipitation as magnesium ammonium phosphate.

Procedure

Dilute 1-2 ml of urine with 25 ml of 0.0066 *M* zinc-EDTA complex in 8 *N* ammoniun hydroxide. Mix with 2 ml of saturated diammonium hydrogen phosphate solution, and dilute to 50 ml. Allow to stand 1 hour with occasional shaking, and use an aliquot of this solution for the analysis. Polarograph after the addition of 2 drops of 0.5 per cent gelatin, a trace of caprylic alcohol, and after expelling the oxygen with a stream of nitrogen.

Calcium in solutions as dilute as 0.001-0.002 *M* may be determined by an amperometric titration after the addition of a small amount of zinc ion, which does not form a complex until all calcium is complexed with EDTA. The zinc ion therefore acts as an amperometric Indicator. This determination of calcium is not affected by magnesium if the latter is present in less than fivefold excess.

Procedure

Add ammonium hydroxide to a sample containing calcium and magnesium until the final concentration is 4-5 *N* in the base, and treat with 0.01 *M* zinc nitrate solution to give a final concenntration of the latter of 0.001-0.002 *M*. Titrate amperometrically with 0.1- 0.01 *M* EDTA at - 1.5 volts in an atmosphere of nitrogen and in the presence of gelatin.

Sympson has developed an amperometric method for the titration of calcium with disodium EDTA. The method involves the reaction between calcium and disodium EDTA with the zincate ion in potassium hydroxide as the indicator ion, and the measurement of the decrease in diffusion current of the zincate after completion of the calcium -EDTA reaction. By titrating 3-11 mg of calcium in an initial volume of 50 ml, the deviation was 0.59 per cent; but with samples containing not greater than 2 mg of calcium with an inital concentration of less than 0.001 *M*, results were low. In moderate amounts magnesium does not interfere. The method compares favourably in accuracy with other titrimetric methods and is more accurate, but less sensitive than optical methods.

Procedure

Use 45-50 ml of a sample containing not less than 2.2 mg of calcium. Add to this solution 10 ml of 0.01 *M* zincate indicator solution. This is prepared by dissolving 0.01 mole of zinc oxide in minimum amount of hydrochloric acid, diluting to approximately 1 liter, and adding I mole of potassium hydroxide and 0.5 mole of potassium chloride. If the sample is dissolved by adding excess acid, adjust the pH of the solution to nearly neutral before adding the zincate solution. Remove oxygen from the solution by bubbling with oxygen-free nitrogen. Then titrate the resulting solution. The potential applied is –1.700 volts vs S.C.E. Take atleast 4 readings both before and after the end point. Bubble the solution with nitrogen for atleast 2 minutes after each addition from the buret.

The indicator is present at a concentration comparable with that of the sample being determined, and the end point of the polarographic analysis is determined by extrapolation.

Zinc is complexed in the zincate to such an extent that the EDTA reacts with calcium until it is completely titrated, and then forms the EDTA complex with zinc, decreasing the diffusion current proportionally to the decrease in its concentration. Reducible metal ions interfere and must be removed. The typical titration curve indicates constant current before the end point, after which a linear decrease occurs.

The optimum hydroxyl ion concentration is 100–140 times that of the zincate ion. A concentration of 0.02 *M* zincate yields a sufficiently large diffusion current for convenient measurement and is sufficiently sensitive as an indicator. Low results are obtained if the concentration of calcium is less than 0.001 *M*. The presence of magnesium reduces the diffusion current of the zincate ion, but the standard deviation for the titration of calcium is not significantly different from the results obtained in the absence of magnesium.

Determination of Cobalt

Souchay and Faucherre report that in the presence of EDTA, cobalt (II) can be oxidized to cobalt (III) and determined polarographically, since the half-wave potential of cobalt (III) is only 0.2 volt. If iron is present it must be precipitated with pyridine. The displacement reaction.

$$MY^{=} + Ca^{++} \rightarrow CaY^{=} + M^{++}$$

in which M^{++} represents the cations of copper, cadmium, zinc, cobalt, and nickel, and $MY^{=}$ their EDTA complexes, has been studied polarographically. The great mobility of zinc and cobalt and the low rate of displacement of nickel makes possible the polarographic determination of zinc and cobalt in the presence of high concentrations of nickel.

The procedure consists in the conversion of all cations concerned into their polarographically inactive complexes by the addition of EDTA, and the displacement of cobalt from its complex by the addition of calcium ions in 4 *N* ammonium hydroxide solution. After the addition of gelatin, the solution is polarographed in the usual manner.

Procedure

To the cobalt solution add equal volumes of EDTA and calcium nitrate of the same molar concentration, make basic with ammonium hydroxide to a final concentration of 4-5 *N*. Add gelatin to suppress the cobalt maximum, and measure polarographically.

Determination of Copper

Pecsok has studied the use of EDTA as a supporting electrolyte for the polarographic determination of copper. Both the reduction of the copper-EDTA complex and the oxidation of copper from a dropping amalgam electrode into an EDTA electrolyte are nearly reversible and yield well-defined polarographic waves. The diffusion curve is proportional to the concentration of copper from 10^{-5} through 10^{-2} *M*. For 0.25 *M* EDTA, at pH 6-9, the diffusion curve constant is 2.85, and the diffusion coefficient is 5.47×10^{-6} sq cm per second.

Faucherre and Souchay have described a polarographic method for the determination of traces of copper in alloys containing large quantities of tin, lead and antimony.

Bennett and Schmidt have studied the polarographic behaviour of the copper – EDTA complex in aqueous sodium sulphate solutions over the pH range 3.5 – 13.5.

Reaction with Europium

Onstott has studied the complexes of europium polarographically. In the presence of a large excess of EDTA, europium (III) is reversible reduced to europium (II) at the droping mercury electrode in the pH range 6–13.6.

Determination of Germanium

Valenta and Zuman have studied the polarography of germanium. Germanium (IV) showed a polarographic wave of –1.4 volts in a solution 0.1 *N* in ammonium hydroxide and 0.1 *N* in ammonium chloride. If the determination of germanium is carried out in 0.1 *M* EDTA at pH 6-8, a wave at –1.3 volts is formed. This wave is practically constant over the pH range 5-9 and is not disturbed by an excess of zinc and arsenic (v). Arsenic (III) gives a wave that is 0.3 volt more negative. Optimum concentrations of germanium for the determination are 5×10^{-5} to 10^{-4} *M*, with a minimum of 5×10^{-6} *M* at pH 7.6-8. The Interfering effect of silica is eliminated by the addition of 10^{-4} *M* fuchsin.

Determination of Iron

Souchay and Funcherre[2] have investigated the polarographic determination of iron with solutions of EDTA. Small quantities of iron may be determined in the presence of copper and other metals by complexing copper with EDTA.

Faucherre and Souchay have described a polarographic method for the determination of traces of iron is alloy containing large quantities of tin, lead and antimony.

Determination of Molybdenum

A polarographic method for the determination of molybdneum in the presence of EDTA is based on the formation of a complex with molybdenum and the complexing agent in a weakly acidic solution. This method is suitable for the determination of traces of molybdenum in tungsten. It may be applied in the presence of copper, lead, and other elements which are reduced at the dropping mercury electrode at potentials more negative than the reduction potential of the molybdeum complex. An acetate buffer consisting of 0.1 *M* acetic acid and 0.1 *M* ammonium aceate is the most suitable support electrolyte. In this medium molybdenum shows a well-developed reduction wave in the presence of EDTA with a half-wave potential of -0.63 volt vs S.C.E. Under these conditions of acidity, the height of the molybdenum wave is directly proportional to the molybdenum concentration and is independent of the amount of EDTA present. This linear relationship holds through the concentration range 10^{-2} to 10 10^{-5} *M*. Under the conditions described, tungsten is polarographically inactive.

Feltham and Martin have used salts of EDTA as the supporting electrolyte and complexing agent in the determination of vanadium (V) and molybdenum(VI). The concentration of the vanadium –EDTA complex is proportional to the difussion current upto a concentration of 1.3×10^{-3} *M* when measured at –160 volts vs the S.C.E., at a pH of 10.0. Molybdenum gives a double wave in acid solution at a pH of 2.8. The half-wave potentials are –0.244 and –0.608 volt vs. The S.C.E. The heights of the waves are proportional to the molybdenum concentration up to a concentration of 1.1×10^{-3} *M* when measured at –0.450 or –0.850 volt vs the S.C.E.

Determination of Nickel

Faucherre and Souchay have described a polarographic method for the determination of traces of hickel in alloys containing large quantities of tin, lead and antimony.

Determination of Niobium

The polarography of a solution of niobium in 0.1 *M* disodium EDTA at pH 3.05 gives waves with half-wave potential of –0.609 volt and –10.5 volts vs S.C.E. The latter wave, coalesces with hydrogen wave, but the former is shown to be reversible at pH 3.2, and to be proportional to the concentration of niobium over the range of 5–500 μg per ml. The sensitivity of this method can be increased by square-wave polarography.

Determination of Thallium

Pribil and Zabransky studied the polarography determination of thallium in the presence of EDTA. Their investigations have shown that thallium does not form a complex with EDTA in neutral or acidic solution, but the formation of a complex in a basic solution, is indicated by a displacement of the half-wave potential of thallium to more negative values. Lead, bismuth, iron (III) also form complexes is acidic solution, and their half-wave potentials are also displaced to more negative values. Differences in half-wave potentials make it possible to distinguish between the waves of copper, thallium and lead. On the basis of these observations, a method has been developed for the determination of as little as 10 μ g of thallium in biological materials.

Transfer 100–200 ml of urine, or 10-20 mg of blood or tissue to a Kjeldahl flask and digest with sulphuric acid with alternate additions of nitric acid and 30 per cent hydrogen peroxide. Continue heating after the addition of hydrogen peroxide until fumes of sulphur trioxide appear, and then cool. Cautiously neutralize the reaction mixture with ammonium hydroxide and add sufficient ammonium hydroxide to bring the pH to 11 (Alizarin GG indicator). The final volume of solution should not exceed 100 ml. Now treat with 0.5 g of potassium cyanide and 0.5 g of ammonium tartrate, and extract with successive portions of 5, 10, 10 and 15 ml of a 0.1 per cent solution of dithizone in chloroform. Thallium, lead, and bismuth pass quantitatively into the organic layer. Evaporate the combined chloroform extracts to dryness, and again heat the residue with 0.5 ml of concentrated sulphuric acid and 30 per cent hydrogen peroxide. When the last traces of peroxide are destroyed, neutralize the mixture with ammonium hydroxide and evaporate to dryness. Take up the residue in 5 ml of 0.1 *M* EDTA solution and 5 ml of acetate buffer of pH 4, and polarograph the solution between -0.3 and -0.8 volt.

Of the commoner elements which may be present, only bismuth causes interference

Heyndrickx and Collaer have studied the polarographic, voltametric, and amperometric characteristics of thallium and lead with EDTA. Thallium (III) is reported to form stable complexes below pH 8.9, but not above this pH. Lead (II) forms stable complexes between pH 4.7 and 11.

Determination of Titanium

Blumer and Kolthoff have studied the polarographic relations of titanium (III) and titanium (IV) in complexes with EDTA. Such complexes, probably TiY – and TiY, are formed at pH 2.5 – 1.0. At this acidity, the half-wave potential of the reduction wave for the titanium complex is –0.22 volt vs the S.C.E. and +0.3 volt vs the hydrogen electrode, and is independent of pH and concentration of the complex. The half-wave potential is similar for the titanium (III) complex. Pecsok and Maverick have also made a polarographic study of the titanium EDTA complexes. Titanium (III) and titanium (IV) in solutions of EDTA yield well-formed, reversible polarographic waves. Below pH 2, the half-wave potential is independent of pH. Above pH 2.5, the dependence is linear, and indicates that 2 hydrogen ions are involved in the electrode reactions. As the pH is increased, the waves become more irreversible, the diffusion curve of the wave is a direct measure of the concentration of the complex in the presence of either an excess of titanium or of the complex former. Measurements indicate that below pH 2, the titanium complex TiY is reduced to the TiY – complex. The TiY complex is in equilibrium with the TiO^{++} ion. Above pH 2.5, $TiOY^{=}$ complex is the predominant species and also reduced to the TiY – ion. As the pH is increased above 3, the TiO^{++} ion becomes increasingly hydrolyzed to colloidal titanium dioxide. Complications arise from the numerous titanium species but in moderate concentrations of EDTA and at pH 2-5, the titanium waves are reproducible and are analytically useful.

The reduction of the titanium –EDTA complex at the dropping mercury electrode occurs reversibly. The half-wave potential with reference to the saturated calomel electrode depends upon the pH and is –0.555 volt at pH 5.5, –0.427 at pH 4.6, -0.418 at pH 3.4 and -0.391 at pH 2.25. At lower pH values, the complex tends to be destroyed. The clearest waves are obtained in 2 *M* sodium acetate and 0.1 *M* EDTA. The diffusion current constant, in milliamperes per millimole per liter for a titanium concentration of 0.234–2.340 milliomole per liter is 3.17 ± 0.11. Over this range of concentration, the relationship between current and concentration is linear. EDTA must be added before sodium acetate, since otherwise titanium forms a complex with acetate that is only slowly destroyed by reacting with EDTA. Iron gives a complex with a half-wave potential in 2 *M* acetate of –0.12 volt and does not interfere unless the concentration is 3–4 times that of titanium. Vanadium and copper, if present, must be removed.

Sinyakova has determined titanium polargraphically as the EDTA complex in sodium acetate solution. The optimum conditions reported for this determination are : a 2 *M* sodium acetate solution which is 0.1 *M* with respect to disodium EDTA; a pH of 4.5–5.0; and a titanium content of 2.34×10^{-4} to 2.34×10^{-3} moles per liter. Under the above conditions the half-wave potential is –0.473 volt. Best results are obtained by dissolving the dissodium EDTA in the sodium acetate solution, and adding to this the titanium solution Iron, vanadium, and copper interfere and must be removed. Chromium is practically without effect on this determination.

Determination of Uranium

Pribil and Blazek have found that the polarographic reduction of uranium (VI) takes place at the same potential in solution containing 10 per cent ammonium carbonate and 0.1 *M* with respect to EDTA as in a solution not containing EDTA. Most other elements form stable complexes under these conditions which are either not reduced at all in the accessible potential range or are reduced at a more negative potential than uranium, yielding limiting currents which are considerably reduced. On the basis of these observations, a polarographic method has been developed for the determination of uranium in the presence of nickel, cobalt, managanese, zinc, chromium, aluminium, beryllium, titanium, and other elements. Only lead and relatively large amounts of copper interfere. This procedure is illustrated by the method for the determination of uranium in auxenite.

Procedure

Fuse 0.2 g of the finely ground sample with 4 times its weight of sodium carbonate in a platinum crucible. Remove silicic acid by filtration, and evaporate the filtrate to dryness. Extract with 10 ml of hot 1 *M* solution of ammonium carbonate. Filter and dilute the filtrate to 25 ml, and then use 5 ml of this solution for a polarographic measurement after the addition of 5 ml of 1 M ammonium carbonate solution and 0.1 g of solid EDTA. The concentration of uranium is determined by the method of standard additions.

Determination of Vanadium

Pecsok and Juvet have studied the polarographic characteristics of vandium complexed with EDTA. Vandium (III) and vandium (II) complexes give single reversible waves at pH 5–12.6. The half-wave potential is independent of pH at 5–8.5 and equal to –1.2 volts vs the S.C.E. At higher pH the potential is independent of pH. The vanadayl complex shows a characteristic pre–wave, believed to be rate controlled reduction of vanadium (IV) to vanadium, (III) followed by the main wave, which corresponds to reduction of vanadium (III) to vanadium (II). For these three oxidation states the half-wave potential is independent of the concentration of the complexing agent. These complexes may be useful for the polarographic determination of vanadium when other ions interfere with the non-complexed wave. Polarograms of the vanadium (III) complex show two waves in acidic solution.

Feltham and Martin have used salts of EDTA as the supporting electrolyte and complexing agent in the polarographic determination of vanadium (V).

Micka and Tockstein have studied the polarographic behaviour of vanadium (VI) and vanadium (V) salts in a borax, sodium carbonate, Britton-Robinson buffer, acetate buffer, and in sulphuric acid. The effects of adding disodium EDTA have also been studied, and the polarographic curves presented.

Determination of Zinc

The replacement of a metal ion by calcium in the reaction

$$MY^{=} + Ca^{++} \rightarrow CaY^{=} + M^{++}$$

can be followed polarographically. The great mobility of the zinc lion and low rate of displacement of nickel makes a polarographic determination of zinc possible in the presence of nickel. The procedure consists in the conversion of all cations concerned into their polarographically inactive complexes by the addition of EDTA, and the displacement of zinc from its complex by the addition of calcium ions in 4 *N* ammonium hydroxide solution. After the addition of gelatin, the solution is polarographed in the usual manner. A high concentration of nickel causes the appearance of a small wave in front of the principal zinc wave.

Faucherre and Souchay have described a polarographic method for the determination of traces of zinc in lubricating oils. The titration is carried out with EDTA at the dropping mercury electrode in a basic soloution containing cyclothexylamine, or in a buffered solution containing ammonium aceate. The method is applied to the determination of zinc in new oils containing barium, and phosphorus, and in used oil containing barium, calcium, iron, copper, lead, and phosphrus. Iron and copper are removed by an either extraction of the cupferron complexes, lead and barium as the sulphates, and calcium suppressed by the addition of sodium fluoride.

10

Amperometric Titrations

General

In any titration process the concentration of the reaction decreases, that of the titrant increases, and those of the products increases as the titration progresses. Where one or more of these substances is capable of reduction or oxidation at a microelectrode, the voltametric or polarographic technique may be used to follow the course of the titration and to provide a means of locating the equivalence point volume. When such a method is used, the titration is called *amperometric titration* and a plot of diffusion current or limiting current vs volume of titrant will yield a curve from which the equivalence point volume may be evaluated. The amperometric technique is subject to less error than corresponding voltametric or polarographic method. With the polarographic titration method, for example, the sup porting electrolyte and capillary characteristics exert less influence, and the temperature, while it should remain constant during titration, is of not such fundamental importance. Moreover the reactant itself need not be reactive at the microelectrode and the method may be based on a reactive titrant or reaction product. The method of plotting the curve involves data accumulated on either side of but well removed from the equivalence point zone, so that titration reactions which are incomplete to a considerable degree at the equivalence point may be adapted to amperometric titration technique. The degree of dilution of the reactant and / or titrant is not important as it is in standard work, since the method is basically suited to low concentration media.

Polarographic Amperometric Titrations

The general forms for the amperometric titration curves where

(*a*) the reactant is DME reactive
(*b*) the titrant is reactive
(*c*) a product of titration reaction is reactive

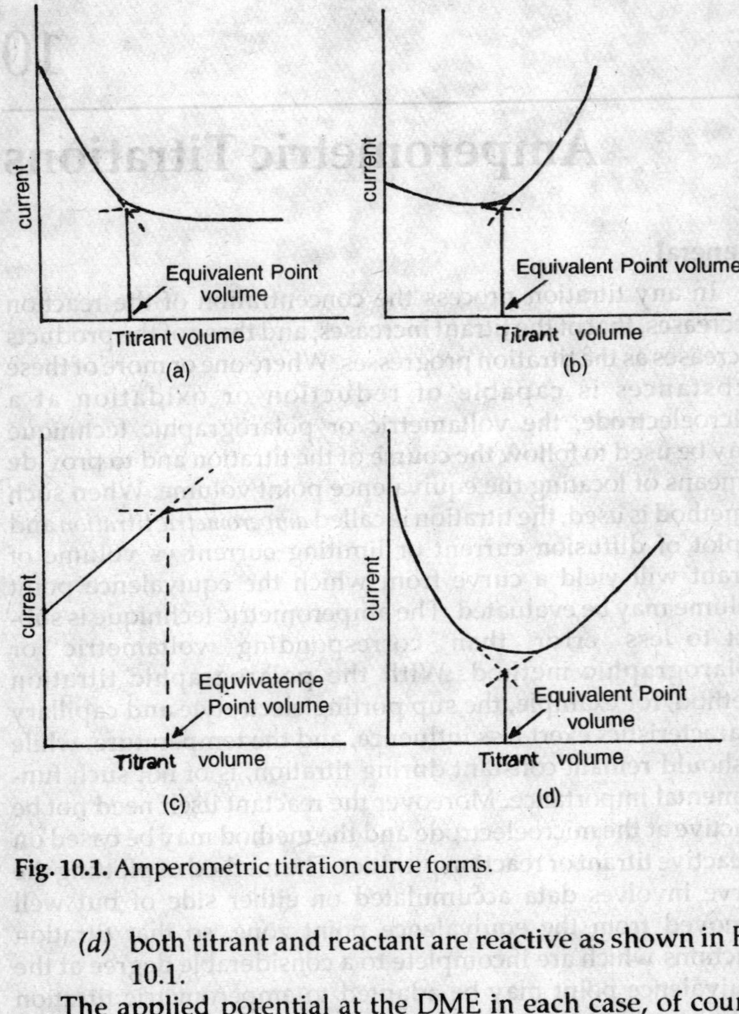

Fig. 10.1. Amperometric titration curve forms.

(*d*) both titrant and reactant are reactive as shown in Fig. 10.1.

The applied potential at the DME in each case, of course, appropriate to the plateau current zone for the electrode reation involved. The equivalance point volume location technique is illustrated for each titration form.

In the amperometric titrations it is necessary to correct each current reading, usually the limiting current value, for the volume dilution effect of the titrant. Here the starting volume of the reactant solution must be known and the correction involves multiplying each current value by the factor $\frac{v + \vartheta}{v}$, where V is the volume (starting and v is the volume of titrant solution added to the particular point in the titration. Alternately the concentration of the titrant solution may be made at least 20 times that of the reactant solution, so that the volume v at any point is always very small compared with V. In this way, the use of the correction factor may be avoided with liltle in the way of error introduced.

Amperometric Titrations with the Rotating Platinum Microelectrode

The DME has the disadvantage of mercury being susceptible to attack by oxidising agents of various tyres and by certain other substances. Where this solution may exist and where the required applied potential does not exceed that at which hydrogren is discharged at platinum (a value apprciably less negative than that for mercury), the rotating platinum electrode may be used. This micro-electrode consists of a very short length of platinum wire fused into the side of a glass tube and connected through a mercury column inside the tube, to the applied potential source. The tube is rotated, without precession, at a constant *rmp* of about 600 by a synchronous motor. Although certain fundamental differencies exist between the solution currents for this system and the polarographic system, the method of equivalence-point equivalence-point volume location is generally similar.

Apparatus Used in Amperometric Titration

To carry out amperometric titrations we make use of either a DME or a rotating platinum electrode. The apparatus used is shown in Fig 10.2.

It consists of an indicator electrode (DME) and a calomel reference electrode. A galvanometer is used to measure current, the rheostal placed in series is used to adjust the senitivity of galvanometer.

To do the amperomertric titration the potential difference between the indicator electrode and reference electrode is kept constant and the value of current *(G)* is noted after each addition of titrant.

If the platinum rotating microelectrode is used an indicator electrode it gives the following advantages over DME.

(i) It is simple to construct

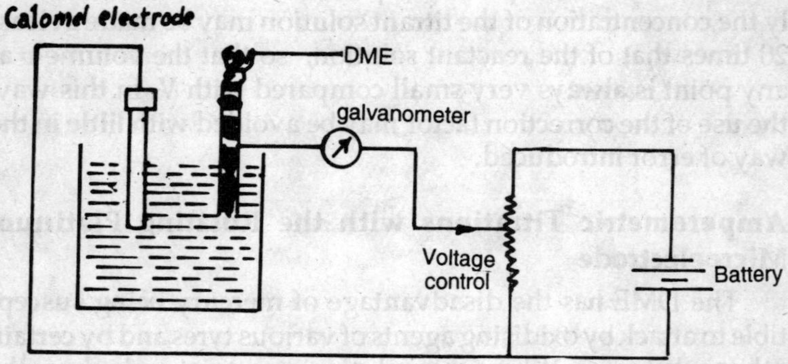

Fig. 10.2

(ii) It is creases the range of the experiment by about 0.9 V an positive side

(iii) The rotation of electrode increases the value of diffusion current by 20 times its value in polarography.

Technique used in Amperometric Titrations

In these titrations oxygen must be removed by bubbling hydrogen or pure nitrogen through solution before starting titration. It is not necessary if the titration is carried out of e.m.f. less than the e.m.f. at which oxygen would have diffusion current.

To illustrate it let us consider the titration of reducible ions such as Pb^{2+} with non-reducible ion like SO_4^{2-} Pb^{2+} being reducible at cathode gives a diffusion current but SO_4^{2-} being

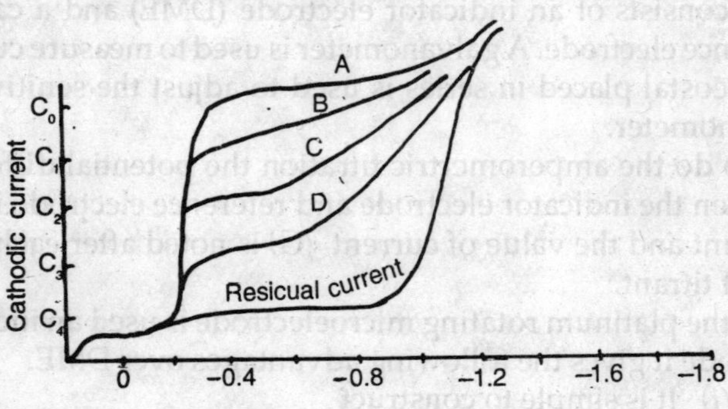

Fig 10.3. Ec vs S.C.E.

non-reducible shows no diffusion current. A polarogram of a solution containing Pb^{2+} is shown in Fig 10.3 (curve A).

At a constant voltage the limiting current corresponds to initial concentration C_o OF Pb^{2+1} ions. The titrant shows no diffusion current at this applied em⁺ when a small amount of SO_4^{2-} is added it removes some of Pb^{2+} ions and the current will decrease as represent by curve B (Fig. 10.3). In the same way we get curves C,D, etc. At i_r, no more of Pb^{2+} ions are left in solution and the current flowing corresponds to the residual current which is characteristic of sup porting electrolyte.

The curve serves two fold purposes:

(*i*) It illustrates the principle of amperometric titrations (current decreases with decrease in concentration of Pb^{2+}.

(*ii*) The plateau of voltage curve gives the region of voltage that should be applied to the system of Pb^{2+} ions titrated amperometrically against SO_4^{2}– ions.

If we plot a curve of diffusion current against volume of titrant in this titration a curve shown in Fig. 10.4 is obtained.

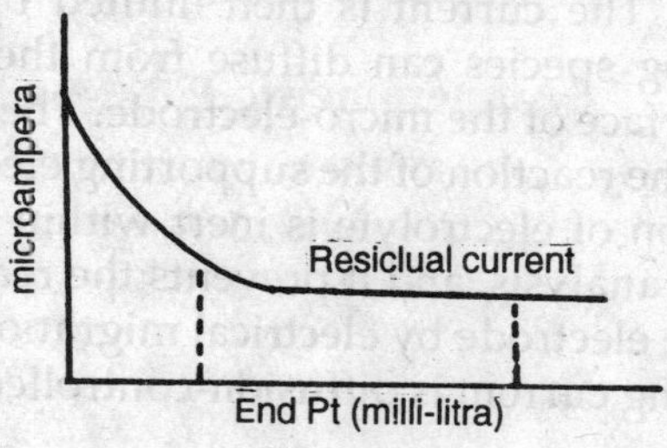

The end point is the point of intersection.

The polarographic may be applied to the detection of the end point of volumetric procedures involving precipitations, complex, formation, neutralisation and oxidation-reduction. As the titration proceeds, the electro reducible substance is removed from the solution and the diffusion current steadily decreases until it reaches a constant value provided the titrant is not electro-reducible or electroxidisable. If the diffusion current is plotted against volume of titrant added, two straight lines are obtained. The point of intersection of these lines represents the end point of the titration.

Since it is the current that is measured in this procedure, it is customary to refer to the method as amperometric titration.

Polarography, as already discussed. The polarogram obtained by the measurement provides qualitative and quantitative information on electro reducible and electro-oxidizable substances. The normal concentration range for substances being analyzed is from 10^{-2} molar to 10^{-5} molar.

In conventional polarography, the micro-electrode is a dropping mercury electrode (D.M.E.) consisting of small reproducible drops of mercury flowing from the orifice of a capillary tube connected to a mercury reservior. A saturated calomel electrode (S.C.E.) with a large surface area is the most commonly employed reference electrode. As the voltage applied to the cell increases, only a very small residual current flows until the substance under assay undergoes reduction or oxidation. Then the current increases, at first gradually, then almost linearly with voltage, and it gradually reaches a limiting value. On the initial rising portion of the polarographic wave, the increased flow of current results in a decrease in the concentration of the electro-active species at the electrode surface. As the voltage and current increase, the concentration of the reactive species decreases further to a minimal value at the electrode surface. The current is then limited by the rate at which the reacting species can diffuse from the bulk of the solution to the surface of the micro-electrode. The final current rise is caused by the reaction of the supporting electrolyte. This large concentration of electrolyte is inert within the potential range used in the analysis, and it prevents the reactive species from reaching the electrode by electrical migration thus assuring that the limiting current is diffusion-controlled.

Advantages to Amperometric Titrations

The amperometric titrations have several advantages over other types of titrations like potentiometric titrations, conductometric titrations etc. The important advantages are:

(i) These titration can be carried out in simple vessels.
(ii) The characteristics of electrodes are not important in amperometric titrations.
(iii) These titrations are least affected by any disturbing factors present in solution.
(iv) These are much more sensitive
(v) It can be adapted to precipitation reactions.
(vi) These titrations can be carried out even by using rotating platinum electrodes.

(vii) It is not essential to keep current constant during titration.

(viii) The reversibibity of a reaction is not essential in these titrations.

(ix) Even the depolarising substances can be accurately estimated amperometrically.

(x) These can be carried out rapidly because end point is obtained graphically.

(xi) These titrations are also possible for such reactions in which substance formed is slightly soluble as in precipitation titration.

Disadvantages of Amperometric Titrations

These titrations have the following draw backs:

(i) Because of coprecipitation and post-precipitation errors are introduced in amperometric titrations.

(ii) When foreign substances (sup porting electrolytes) are present there is no change or very little change in current during amperometric titrations.

Biamperometric Titrations

As a modification of amperometric titrations, two stationary polarizable microelectrodes are immersed in a sample solution. A small potential difference in applied between the two electrodes and the current that flows is recorded as a function of the volume of added reagent. The end point is marked by

(a) a sudden decrease in the current to zero level followed by rise in current

(b) a sudden rise in current from zero level or

(c) a sudden approach to zero level current.

Previous by it was known as dead-stop end point method and is now know as "biamperometric titration." Such titrations are applicable only in those cases where oxidation-reduction system may be present both before and after the end point.

Titrations of $KMnO_4$ against type may be carried out in this way.

The sensitivity of biamperometric titrations is dependent on the electrode area, applied voltage and speed of stirring the solution. The most commonly used electrode materials are platinum, gold, mercury, copper, graphite and antimony.

These titrations are most applicable in series analysis because of their simplicity, speed and suitability for automation.

In addition to the use of these titration for redox system, Iodometry etc. They are also used for the determination of water with Karl Fischer reagent. It is also possible to follow reactions in which water is liberated (esterification) or consumed (hydrolysis).

11

Voltammetry

In voltammetry an electron-transfer reaction can take place at the electrode if the potential is appropriate, its extent being determined by the surface concentration of some electroactive species. The resulting current will be only a transient, decaying rapidly to zero, unless some mechanism is present to bring a continuously renewed supply of the electroactive material to the surface. A similar mechanism must generally be available to remove the product of reaction from the surface.

Hence, the subject of voltammetry must treat, not only the electron-transfer process itself, but also the detailed mechanisms of mass transport. This requires knowledge of the geometry of the cell and of the working electrode. In this section we explore the interdependence of these factors, and subsequently show how they can be applied in a practical sense in polarographic and other analytical techniques.

Two major modes of transport can be distinguished, due to: (1) convective motion of the solvent (and supporting electrolyte), carrying the active species with it, and (2) movement of the active species through the solvent, as by diffusion. Both of these modes are acting all the time that a current is passing through a cell, but usually one mode predominates.

In practical terms, two kinds of voltammetric experiments are commonly encountered, conducted with or without mechanical stirring. If the solution is stirred, clearly the resulting bulk motion of the liquid is the dominant transport mechanism, but even here diffusion plays a part. It is customary to hypothesize a thin stationary layer of solution in contact with the electrode, that is unaffected by stirring; the mechanism of transport across this thin layer is by diffusion.

In the other limiting case, the cell is protected as far as possible from mechanical forces that might cause convective motion of the solution. Here, the transport of active species is by *diffusion,* in which the motion is caused by a gradient in chemical potential, or by *migration,* the effect of an electrical potential gradient. Even in this circumstance, however, mechanical convection is never entirely absent, the residual effect being due to such sources as temperature changes and vibration. In some situations, the faradaic effects of the current may produce a solute of different density, and this causes some convection. The growth and fall of the mercury drops in polarography also creates a small convective effect.

Of all these mass transport mechanisms, diffusion and migration are most readily susceptible to detailed theoretical treatment. In the presence of supporting electrolyte, migration becomes of negligible importance.

It is instructive to consider the nature of the current response to a step potential function in the two limiting cases previously discussed. The corresponding current-time curves are plotted in Figure 11.1. The working electrode is specified only in that it is chemically inert and of fixed area. Zero time is taken as the instant when the potential is changed from well below to well above the formal potential $E^{o\prime}$ for an electroactive species. In an unstired solution, the surface concentrations change rapidly in the sense necessary to diminish the current. This is given by the Cottrell equation previously described:

$$I = nFA\,(D/\pi t)^{1/2}\;C^{*} \qquad (11.1)$$

Thus, the current drops off as the inverse square root of the time. (The symbol C^{*} represents the bulk concentration of the active species).

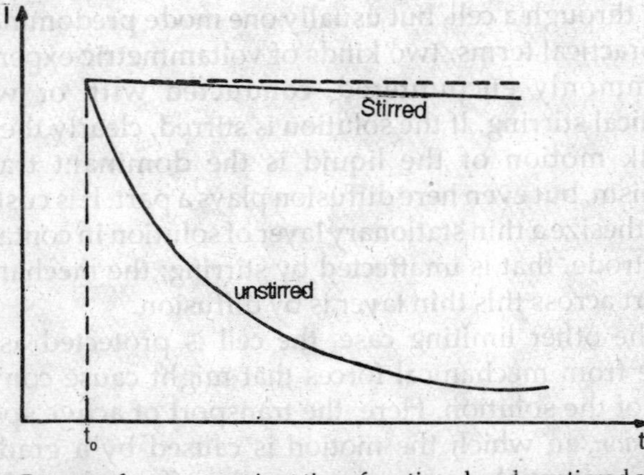

Fig. 11.1. Curves of current against time for stirred and unstirred solutions. The slope of the upper curve is exaggerated for clarity.

In the case of a well-stirred soluton, fresh active material is always available to the electrode, so the current stays at a nearly constant value. The magnitude of the current is proportional to concentration, which diminishes slowly because of the charge-transfer reaction, and the curve drops exponentially with time at a slow rate.

In the absence of stirring, there is no noticeable change of bulk concentration, simply because during all but the first small fraction of the time, the current is very small (Figure 11.1). By contrast, if the solution is stirred, the current continues at a higher level, and appreciable changes in concentration may ensue.

One of the most useful electrodes for the study of voltammetry in unstirred solutions is the *dropping mercury electrode* (DME), which consists of a continuing series of minute droplets of mercury issuing under pressure from a glass capillary. Voltammetry at the DME, called *polarography,* is the principal subject of the present chapter.

Voltammetric methods can, in principle, be approached in either of two ways: the *current* can be recorded as the voltage is controlled, or alternatively, the *voltage* is monitored as a current is impressed. This complementary relation suggests that the phenomena involved are, in essence, concerned with the *impedance* between the electrode and the solution.

Linear Sweep

It is convenient to distinguish two types of voltammetric experiments which are respectively, dependent on or independent of, the scan rate. The various types of polarography belong to the second category, since the results of the measurement do not depend on how fast the voltage range is scanned. One could even jump at random between voltage without changing the current-voltage relationships, as long as the voltage is constant during the life of each mercury drop.

In contrast, in the method to be described in this chapter, the measurements *do depend* on the rate at which the voltage is scanned. The current observed also depends on the exact form of the scan, which is usually a simple linear sweep.

Rate-dependent voltammograms are obtained with solid electrodes or stationary mercury drops, rather than with the DME. In spite of this, the term *stationary-electrode polarography* has sometimes been used to describe this method. Some authors argue that the method is, strictly speaking, neither polarography nor voltammetry, and that it properly should be called *chronoamperometry with linear potential sweep*, a wounderfully descriptive but too lengthly title. We shall use the name *linear sweep voltammetry (LSV)*. The term *cyclic voltammetry (CV)* describes a variant in which the potential is scanned back and forth between two limits. The method in its modern form owes much to the efforts of Shain and Nicholson.

The Basic Experiment

Consider the instrument of Figure 11.2 in which a signal generator produces a voltage sweep from E_i to E_f. A potentiostat applies if to the electrode, which can be a solid such as carbon or platinum, or a stationary mercury drop. The SMDE electrode, can be used conveniently.

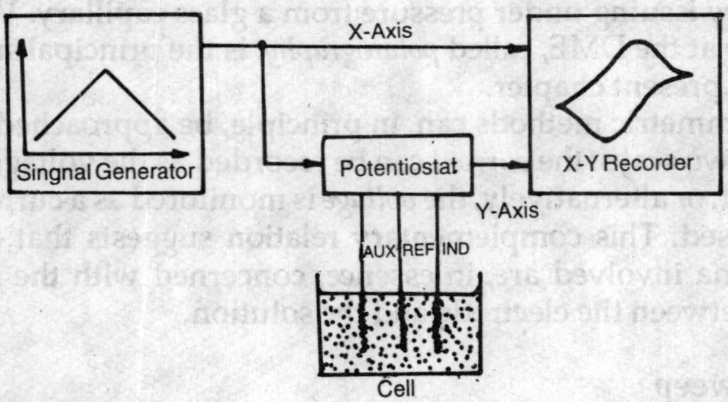

Fig. 11.2 Block diagram of an instrument for cyclic voltammetry.

Experiment

An example of the resulting current for a reversible system is shown in Figure 11.3. A characteristic peaked curve is obtained, occuring in the region of the polarographic wave. Note that $E_{1/2}$, as defined in polarography, falls about midway between the half-peak potential $E_{pk/2}$, and the peak itself, E_{pk}. The value $E_{1/2}$ can be estimated from a linear sweep polarogram, even through it is not directly related to it.

The descending part of the curve follows a $1/\sqrt{t}$ dependence, so that for fast scans the current has a shorter time in which to decrease, and the curve decays less than in a slower scan. The

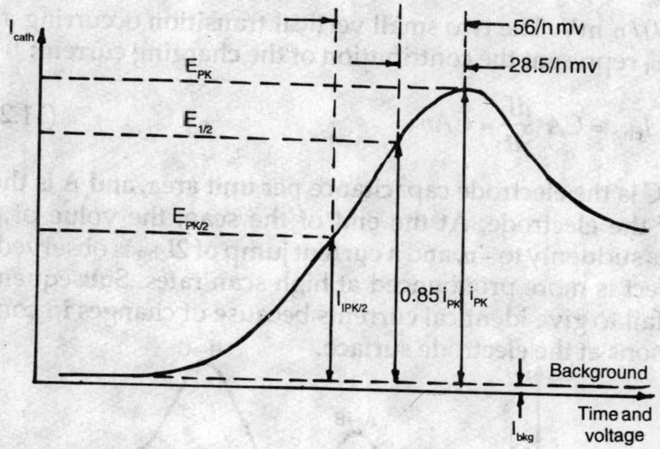

Fig. 11.3. A typical linear-sweep voltammogram for a reversible system. The potential differences indicated are valid at 25°C. The symbol i_{bkg} denotes the residual current background at E_{pk}. The quantity $E_{pk/2}$ is defined as the potential where $I = I_{pk}$.

peak height changes proportionally to the square root of the scan rate, *v*, as well as to the concentration of the active species.

For cyclic voltammetry, the scan is continued in the reverse direction from E_f back to E_i, and a downward-going curve results. The usual practice is to present it inverted on the same graph, as illustrated in Figure 11.4 by plotting against voltage rather than time.

The two peaks in Figure 11.4, describing the reduction (upper peak) and oxidation (lower peak), are separated by

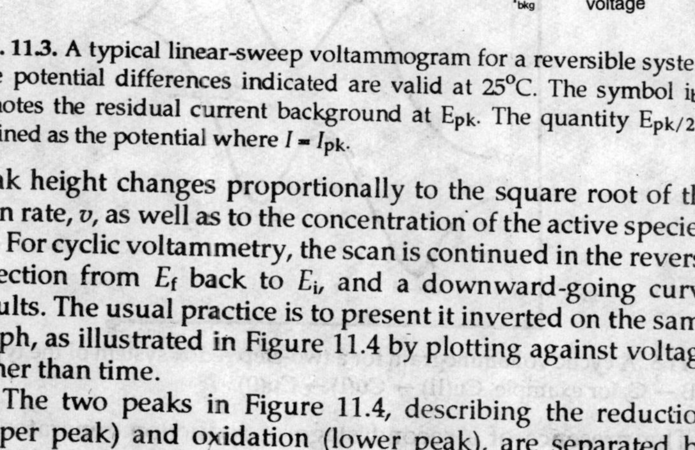

Fig. 11.4. A cyclic voltammogram for a reversible system.

about $60/n$ mV. The two small vertical transition occurring at E_i and E_f represent the contribution of the charging current:

$$I_{chg} = CA\ \frac{dF}{dt} = CAv \tag{11.2}$$

where C is the electrode capacitance per unit area, and A is the area of the electrode. At the end of the scan, the volue of v changes suddenly to $-v$, and a current jump of $2I_{chg}$ is observed. The effect is more pronounced at high scan rates. Subsequent cycles fail to give identical currents because of changes in concentrations at the electrode surface.

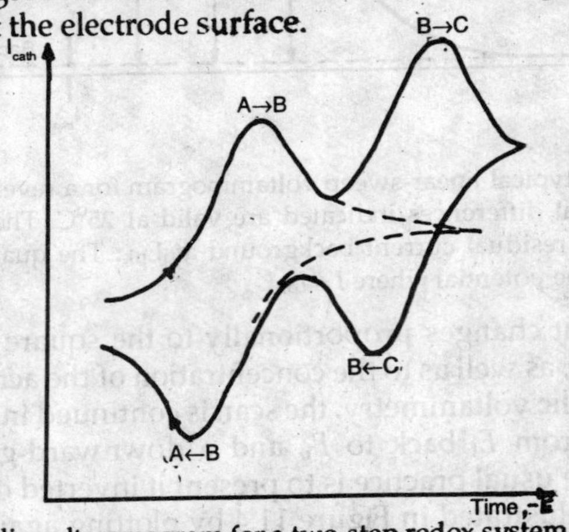

Fig. 11.5. A cyclic voltammogram for a two-step redox system of the type A-- B → C, for example, Cu(II) → Cu(I) → Cu(0).

The presence of a second stage of reduction generates a second pair of peaks, as shown in Figure 11.5. The dashed line represents the curve that would have been obtained in the absence of the second peak and constitutes its baseline. Unfortunately compensation for the base is not straightforward in this case, but modern computer techniques permit such deconvolution, as reported by Gutknecht and Perone.

Theoretical Considerations

In this section we will attempt to show the relation between a voltage sweep stimpulus and the electrode response. The interplay between diffusion and electron-transfer kinetics will be examined in some details.

The current in a voltammetric experiment is, in general, the resultant of serveral components:

$$I_{tot} = I_{far} + I_{chg} + I_{ads} + I_{cross} \qquad (11.3)$$

which describe the contributions of the faradaic, charging, and adsorption currents, as well as a cross term arising from the interaction between the first three. Only the faradaic current is of analytical interest, and the other terms can usually be considered as noise or background. The resolution, sensitivity, and general usefulness of voltammetry depend, in large measure, on the possibility of minimizing or correcting for such background contrbutions.

The charging current in LSV is given by:

$$I_{chg} = \frac{dQ}{dt} = \frac{d}{dt}(CAE) = CA\frac{\partial E}{\partial t} + AE\frac{\partial C}{\partial t} \qquad (11.4)$$

where Q is the charge on the electrode, A the area (which is constant), C the capacitance per unit area, and E the voltage relative to the PZC. The term in $\partial C/\partial t$ arises from the fact that, as the scan proceeds, the capacitance of the electrode changes. Both terms depend on the scan rate $v = dE/dt$.

An adsorption contribution is present if the amount of charged species adsorbed varies during the scan. The effect can be written as:

$$I_{ads} = nFA\frac{d\Gamma}{dt} \qquad (11.5)$$

where the symbol Γ denotes the surface excess of the particular charged species involved. At higher scan rates, $d\Gamma/dt$ is larger. The current may be of either sign, depending on whether cations or anions are adsorbed or desorbed.

The background current can be substracted using a separate blank run, but the correction is valid only if $I_{cross} = 0$. The presence of a cross term signifies that the faradaic current has some effect upon the background current, so that the correction procedure is vitiated. Hence a baseline derived by extrapolation from the curve itself may be preferable.

The faradaic current is the result of the requirement that the ratio of OX/RED be altered when the potential is shifted. The Nernst equation, for a reversible system, can be solved for the ratio of surface concentrations:

$$\frac{C^S_{OX}}{C^S_{RED}} = \exp\left\{\frac{nF}{RT}(E - E^{o'})\right\} \qquad (11.6)$$

Irreversible processes do not obey this equation.

In a LSV experiment, the potentiostat sweeps the applied voltage from an initial value E_i, at rate v, so that $E = E_i + vt$, and the concentration ratio changes with time:

$$\frac{C^S_{OX}}{C^S_{RED}} = \exp\left\{\frac{nF}{RT}(E_i - E^{o\prime} - vt)\right\}$$

$$= \left(\exp\left\{\frac{nF}{RT}(E_i - E^{o\prime})\right\}\cdot\exp\left\{-\frac{nF}{RT}vt\right\}\right) \qquad (11.7)$$

or

$$\frac{C^S_{OX}}{C^S_{RED}} = g.e^{-mt} \qquad (11.8)$$

where g and m are constants defined by reference to Eq. (11.7). This is shown in Figure 11.6 a.

Consequently, as the voltage is swept, OX wil be converted into RED, as can be seen in Figure 11.6 *b*. The two limiting concentrations are shown unequal, contrary to what one might expect in terms of conservation of mass. In the example shown, the difference is caused by the rapid diffusion of OX toward the electrode. With the transformation into the reduced form, the diffusion coefficient becomes smaller, so that RED does not move away as fast. The electrode process eventually reaches a dynamic equilibrium:

$$\frac{C^S_{OX}\ (\text{anodic limit})}{C^S_{RED}\ (\text{cathodic limit})} = \left(\frac{D_{RED}}{D_{OX}}\right)^{1/2} \qquad (11.9)$$

This effect is often negligible but it is important if the diffusion coefficients are widely different. One of the consequences is that, if for example we start with a solution of 0.01 *M* OX and run a cyclic voltammogram, there is no guarantee that at the start of the reverse scan the region around the electrode will be 0.01 *M* in RED. The square-root term in Eq. 11.8) parallels a similar correction term in polarography, but here the effect is more important.

The changes produced at the electrode by the potential sweep extend into the solution to a depth δ, the *diffusion layer*, that is commensurate with $2\sqrt{Dt}$. Beyond about $4\sqrt{Dt}$, the solution remains essentially unchanged For example, in a rather lengthy scan lasting 1 min, the thickness is less than 1 mm, which can be considered an upper limit. The electrode senses only the concentration C^S at its surface, while the current is

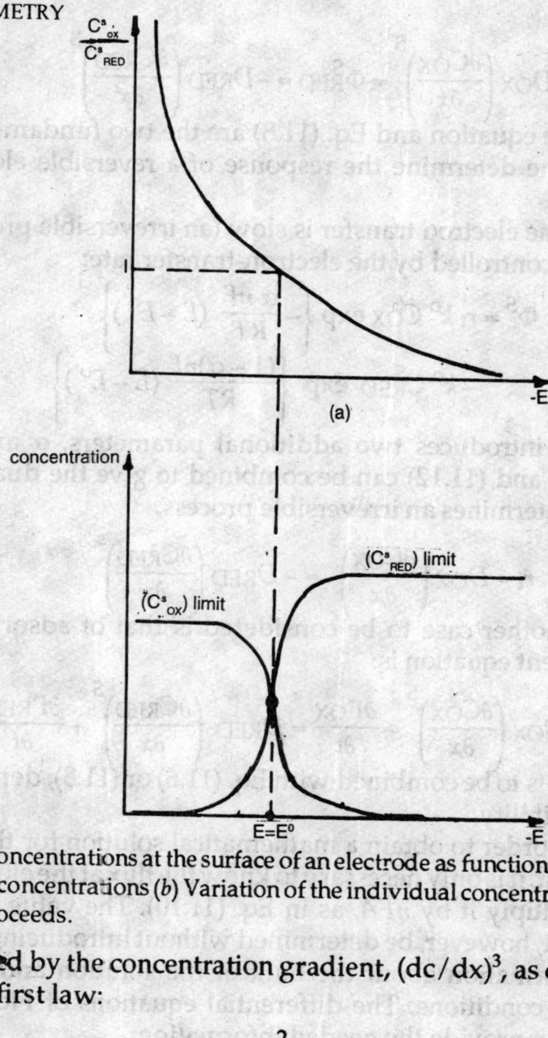

Fig. 11.6. Concentrations at the surface of an electrode as functions of time. (*a*) Ratio of concentrations (*b*) Variation of the individual concentrations as the scan proceeds.

determined by the concentration gradient, $(dc/dx)^3$, as defined by Fick's first law:

$$I = nFA\,\Phi^S\,nFAD_{OX}\left(\frac{\partial C}{\partial x}\right)^2 \qquad (11.10)$$

where Φ^S is the flux of active species at the electrode suface.†

In general, if there is no adsorption involved, the fluxes of RED and OX are much larger than the rate at which they accumulate at the electrode surface. Thus, one can equate the two fluxes:

---

†One can think of a voltammogram as a plot of $(\partial C/\partial x)^S$ vs. time, since this is the only solution property that is actually measured.

$$\Phi^{S}_{OX} = D_{OX}\left(\frac{\partial C_{OX}}{\partial x}\right)^{S} = \Phi^{S}_{RED} = -D_{RED}\left(\frac{\partial C_{RED}}{\partial x}\right)^{S} \tag{11.11}$$

The equation and Eq. (11.8) are the two fundamental relations the determine the response of a reversible electrode in LSV.

If the electron transfer is slow (an irreversible process), the flux is controlled by the electron-transfer rate:

$$\Phi^{S} = r_t k^{o} C^{S}_{OX} \exp\left\{-\frac{\alpha nF}{RT}(E - E^{o'})\right\} - k^{o} C^{S}_{RED} \exp\left\{\frac{(1-\alpha)nF}{RT}(E - E^{o'})\right\} \tag{11.12}$$

which introduces two additional parameters, α and k^{o}. Eq. (11.11) and (11.12) can be combined to give the dual equality that determines an irreversible process:

$$r_t = D_{OX}\left(\frac{\partial C_{OX}}{\partial x}\right)^{S} = -D_{RED}\left(\frac{\partial C_{RED}}{\partial x}\right)^{s} \tag{11.13}$$

Another case to be considered is that of adsorption. The pertinent equation is:

$$D_{OX}\left(\frac{\partial C_{OX}}{\partial x}\right)^{S} - \frac{\partial \Gamma_{OX}}{\partial t} = D_{RED}\left(\frac{\partial C_{RED}}{\partial x}\right)^{S} + \frac{\partial \Gamma_{RED}}{\partial t} \tag{11.14}$$

which is to be combined with Eq. (11.6) or (11.8), depending on reversibility.

In order to obtain a mathematical solution for the faradaic current, it is only necessary to know the flux at the electrode and to multiply it by nFA, as in Eq. (11.10). The value of the flux cannot, however, be determined without introducing additional information about the bulk of the solution and about the initial conditions. The differential equations of Fick's second law can provide the needed information:

$$\frac{\partial C_{OX}}{\partial t} = D_{OX}\left(\frac{\partial^2 C_{OX}}{\partial x^2}\right) \tag{11.15}$$

$$\frac{\partial C_{RED}}{\partial t} = D_{RED}\left(\frac{\partial^2 C_{RED}}{\partial x^2}\right) \tag{11.16}$$

The initial state of the system is given by:

$$\left\{\begin{matrix} C_{OX} = C^{\infty}_{OX} \\ C_{RED} = C^{\infty}_{RED} \end{matrix}\right\} \text{(all } x, t = 0) \tag{11.17}$$

The assumption of semi-infinite diffusion provides two additional boundary conditions.

$$\left.\begin{cases} C_{OX} = C^{*}_{OX} \\ C_{RED} = C^{8}_{RED} \end{cases}\right\} (x \rightarrow \infty, \textit{all } t) \tag{11.18}$$

where the asterisks indicate bulk concentrations. "Infinite distance," for LSV, represents perhaps one milimeter.

The left sides of Eqs. (11.15) and (11.16) represent the time variation of the concentrations. Modifications must be made if homogeneous chemical reactions occur that involve either OX or RED. For example, if RED is involved in a first-order reaction that generates an electrochemically inactive product, the equations become:

$$\frac{\partial C_{OX}}{\partial t} = D_{OX}\left(\frac{\partial^2 C_{OX}}{\partial x^2}\right) \tag{11.19}$$

$$\frac{\partial C_{RED}}{\partial t} = D_{RED}\left(\frac{\partial^2 C_{RED}}{\partial x^2}\right) - k_{hom}\, C_{RED} \tag{11.20}$$

which reflect the effect of the homogeneous reaction rate constant k_{hom}. The above equations describe phenomena occurring within the diffusion layer.

The complete mathematical treatment involves two surtace conditions, such as Eqs. (11.6) and (11.11), and two differential equations, such as Eqs. (11.15) and (11.16), as well as the accompanying boundary conditions. The solution is not simple, because of the exponential dependence on time* for both reversible and irreversible systems. The original work of Randles and Sevcik permits the calculation of the peak current i_{pk} by means of their well-known equations:

$$i_{pk} = 269\, n^{3/2}\, A D^{1/2}\, v^{1/2}\, C^{*}_{OX} \tag{11.21}$$

where a 3/2 - power dependence on the number of electrons is involved. For the numerical constant given, the diffusion coef-

*Considerable mathematical simplification might be obtained by scanning the voltage logarithmically rather than linearly, using a function of the form $E = X + Y \log t$, where X and Y are constants. In this case, Eq. (11.7) would reduce to :

$$C^{S}_{OX} / C^{S}_{RED} = (\text{constant}) \cdot (1/t) \tag{11.20}$$

ficient must be in $cm^2 s^{-1}$, the area in cm^2, and the concentration in moles/liter. The equation can be used to calculate the concentrations when all other quantities are known, but analytically it is preferable to follow a calibration procedure. Good accuracy is obtained over the range 10^{-3} to 10^{-6} M.

The expression for the entire voltammetric curve, as opposed to the value at the peak, is more difficult to obtain, but several numerical calculations have been published, including the classical paper by Nicholson and Shain, from which much of the present discussion is taken.

Their expression for the reversible case consists of an explicit portion and a tabulated *current function*, χ_{rew}:

$$I = 602\, n^{3/2}\, A D^{1/2} v^{1/2}\, C^{*}_{OX}\, \chi_{rev} \qquad (11.22)$$

where we have replaced the symbol for the Nicholson-Shain current function $[\pi\chi(at)]$ by χ_{rev}. A shortened tabulation of values of χ_{rev} is given in Table 11.1. For more details, consult. The symbol, $E_{1/2}$ is here defined as $E^{o'} + (0.0592/n) \log D_{red}/D_{ox})^{1/2}$ just as in polarography. The table shows the peak to occur at 28.5 mV beyond $E_{1/2}$ from which we may deduce that the anodic and cathodic peaks for the same reversible couple are separated by $(2 \times 28.5/n) = 57/n$ mV.

At the half-wave potential, the function χ_{rev} has the value 0.38, which is 85 per cent of the peak value, 0.446, and thus corresponds to 85 per cent of i_{pk}. This suggests a convenient way of calculating $E_{1/2}$ from an LSV curve (cf. Figure 11.3).

Note that the half-wave potential for voltammetry at a fixed-area electrode cannot be defined as the potential where the current is half its maximum value. Nevertheless, in both polarography and LSV, the half-wave potential is a fundamental point of the curve. In LSV, however, the curve is asymmetric, giving a peak, as seen in Figure 11.3.

Another useful parameter of the curve is the half-peak potential, $E_{pk/2}$. This can be measured with precision, whereas the peak potential E_{pk} itself is difficult to measure because of the typically broad summit. E_{pk} can be calculated from $E_{pk/2}$ by the relation:

TABLE 11.1
Values of χ_{rev} at Various Potentials (Relatie to $E_{1/2}$)

$(E-E_{1/2})$ (mV)	χ_{rev}	(mV)	χ_{rev}
+120	0.01	–10	0.42
100	0.02	–20	0.44
80	0.04	–28.5[a]	0.446
60	0.08	–30	0.44
50	0.12	–40	0.44
40	0.16	–50	0.42
30	0.21	–60	0.40
20	0.27	–80	0.35
10	0.33	–100	0.31
0	0.38	–120	0.28

[a]maximum.

$$E_{pk} = E_{pk/2} \pm \frac{56.5}{n} \text{ mV} \qquad (11.23)$$

The negative sign applies to anodic peaks. This equations is sometimes written is the equivalent form:

$$E_{pk} = E_{pk/2} \pm 2.2 \frac{RT}{nF} \text{ mV} \qquad (11.24)$$

Consult Figure 11.3 for a graphical relation between these quantities.

In the case of a cyclic voltammogram, the peaks are separated by exactly $57/n$

TABLE 11.2
Peak Separation for a One-Electron Process [3] (mV)

E_f-E_{pk}	$E_{pk\,(anod)}$ - $E_{pk\,(cath)}$
75	61.8
100	60.5
200	58.3
300	57.8
(Limit)	57.0

mV only if the forward scan caused a sufficiently extensive change from OX to RED. This can be ensured by using a wide potential sweep, or preferably by pausing at E_f for a minute or so before starting the reverse scan. In case neither is done, the reverse peak will be slightly shifted as shown in Table 11.2. From these data, it appears than $59/n$ is probably a better estimate than the theoretical 57/n for reversible peak sepration.

Irreversible Processes

If the electron-transfer process is slow, the differential equations contain the heter ogeneous rate constant k^o and the transfer coefficient α. The solution of the equations involving this condition, as given by Nicholson and Shain is:

$$I = 602\, n(\alpha n_\alpha)^{1/2}\, AD^{1/2}\, v^{1/2}\, C^*_{OX}\, \chi_{irrev} \qquad (11.25)$$

where again a reduction is assumed. Note the unusual replacement of $n^{3/2}$ by $n\,(\alpha n_\alpha)^{1/2}$, where $n\alpha$ denotes the number of electrons involved in the charge transfer step, a rather inaccessible quantity, difficult to separate from α experimentally.

The values of x_{irrev} are given by Nicholson and Shain. The maximum of the curve occurs at a value χ_{irrev} = 0.496. If we assume α to be 0.5, we can calculate the ratio of reversible to irreversible peaks to be:

$$\frac{i_{pk(rev)}}{i_{pk}(irrev)} = \frac{0.446}{\sqrt{0.5 \times 0.496}} = 1.27 \qquad (11.26)$$

so that, everything else being constant, a reversible peak is some 30 per cent higher than an irreversible one.

The irreversible peak current suffers displacement away from $E_{1/2}$ that result from two factors: small values of both k^o and of α. The basic formula describing this displacement is.

$$E_{pk} = E^{o'} - \frac{RT}{\alpha n_\alpha F}\left(0.77 \ln \sqrt{Db} - \ln k^o + \frac{\ln b}{2}\right) \quad (11.27)$$

where $b = \alpha n_\alpha \ Fv/RT$.

By using this equation it is possible to calculate the kinetic parameters αn_α and k^o. A careful examination of the equation indicates that the contribution of k^o to the peak displacement can be compensated by appropriate changes in the scan rate v. Thus, no matter how large k^o may be, that is, no matter how reversible the reaction may be, one can make it behave irreversible by choosing a sufficiently fast scan rate. Conversely, any reaction, at least in principle, can be made reversible by sweeping slowly enough.

There comments underline the fact that electrochemical reversibility is a property of the experimental technique as much as of the chemical system involved.

Coupled Chemical Relactions

If either species OX or RED is involved in a side reaction, an appropriate contribution to the differential equations must be included. The results in each case contain the rate constants of the chemical reactions, and, in addition, k^o and αn_α if the process is irreversible. A few of the many possible case are the following:

$$Z \overset{k}{\rightleftarrows} OX + ne^- \rightarrow RED \ (CE) \quad (11.28)$$

$$OX + ne^- \rightarrow RED \overset{k}{\rightleftarrows} Z \ (EC) \quad (11.29)$$

$$Z_1 \overset{k_1}{\rightleftarrows} OX + ne^- \rightarrow RED \overset{k_2}{\rightleftarrows} Z_2 \ (ECE) \quad (11.30)$$

The symbols on the right indicate the sequence of chemical and electrochemical steps involved. Such reactions have been thoroughly investigated but the details are beyond the scope of our discussion.

The measurement of i_{pk} necessitates a knowledge of the baseline. The problem is especially important on the return scan. If the mechanistic requirements permit, it might be advisable to pause at the end of the forward scan until the current decays to a low value. The effect is illustrated in Figure 11.7. A

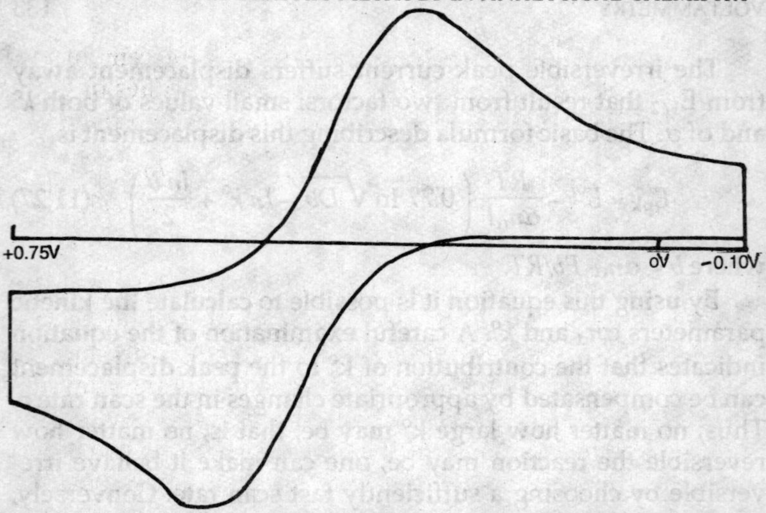

Fig. 11.7. A cyclic voltammogram, showing the drop in current caused by a pause at -0.10 V and again at + 0.75 V.

less desirable alternative is to make a separate experiment recording the current as a function of time after stopping the voltage sweep, as shown in Figure 11.8. The mirror image of this curve (dotted line) is then taken as the baseline for the reverse scan.

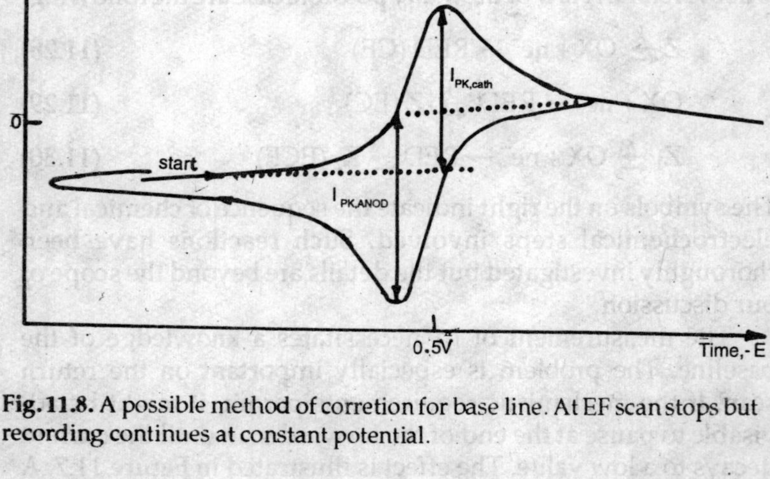

Fig. 11.8. A possible method of corretion for base line. At EF scan stops but recording continues at constant potential.

Staircase LSV

A variant of LSV, which is particularly appropriate for computer control, is stair-case voltammetry, wherein the potential is advanced in steps of δE volts every θ seconds. This corresponds to a voltage ramp of $\delta E/\theta$, the equivalent of the quantity v. Any desired value of the ratio $\delta E/\theta$ can be implemented by various combinations of the two variables. This is a complication in analytical work, but can be useful for mechanistic studies. The sensitivity is good (10^{-7} M) due to the elimination of the charging current.

AC and Pulse LSV

The combination of LSV with AC or pulse signals offers the usual advantages of modulation. The AC method has received attention both in the fundamental and second-harmonic modes. The results for reversible reactions are somewhat surprising, since, at least for slow scans, the process turns out to the time-independent. On the other hand, irreversible systems given time-dependent results that can serve to elucidate kinetic parameters. Pulse voltammetry, especially differential, also gives unique curves. The method seems to promise good analytical results.

12

Finite Diffusion

The majority of electronalytical methods utilize a relatively large volume of solution, extending to some distance from the electrode. If diffusion the controlling transport mechanism, a concentration profile will be formed in the solution as the reaction proceeds. As seen previously, the thickness of the diffusion layer is well below 1 mm. Because of the thinness of this layer, the walls of the cell can be considered to be at infinity. This mode of transport is called *semi-infinite diffusion.* (The "semi" refers to the fact that the process extends in one direction only).

Another type of behaviour occurs when the diffusion is limited rather than infinite. In this event, the response is quite different, even though the excitation may be conventional. The most important possibilities are : (1) the reagent is confined to a layer of solution limited by an inert wall or by another electrode. This can be called *thin-layer bounded diffusion.* It is implemented with so-called thin-layer cells. (2) The diffusion may be confined to a porous memberane in contact with a stirred solution. This is *thin-layer non-bounded diffusion.* Finally, (3), the reagent may be immobilized by adsorption or binding at the electrode surface, in which case only very slight diffusion (or none at all) is needed for the reaction to proceed. This can be called an *immobilized reagent.*

Thin-layer Cells

Although electrolysis with microliter volumes had been attempted as early as 1955, only recently has a thorough investigation been made of electrolysis under conditions where the thickness of the cell is smaller than the diffusion layer. The exact shape of the concentration profile varies with the type of potential time curve used. In order to compare the infinite and bounded diffusion cases, let us first consider the application of a large voltage step to a cell with semi-infinite diffusion. This is

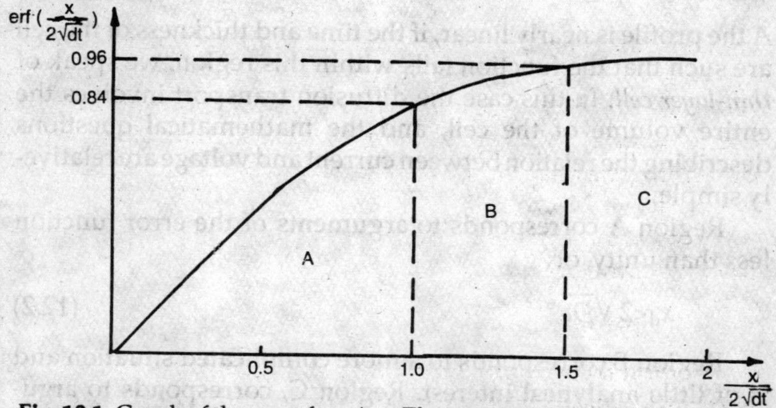

Fig. 12.1. Graph of the error function. The argument of "erf" represents the distance into the solution, in unit of $2\sqrt{Dt}$ cm.

Cottrell-type procedure, and it can be shown that the concentration profile follows a functional dependene of the type:

$$C = C^{*}\text{erf}\,[x/(2\sqrt{Dt}]\qquad(12.1)$$

where "erf" is a tabulated function, the *error function**

The concentration depends on both the distance x from the electrode and the time t. A graph of the spatial dependence at a specified time, is shown in Figure 12.1. As the time increases, the concentration profile becomes shallower, as seen Figure 12.2. In Figure 12.1, three regions can be distinguished. In region

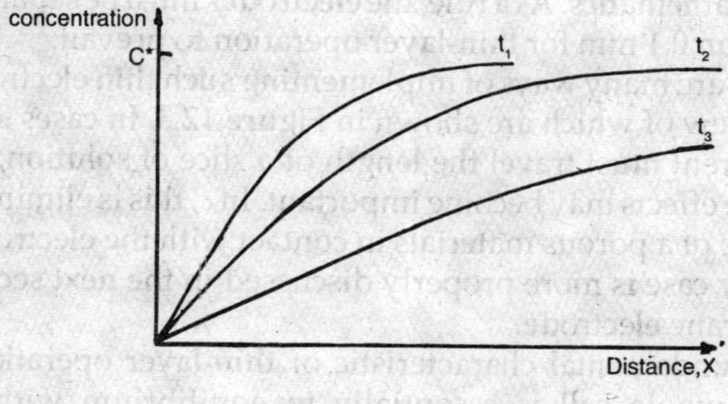

Fig. 12..2. An alternative plot of Eq. (12.1) as concentration profiles at various times. (Curves 1, 2 and 3 correspond to increasing time intervals. The initial slopes of the curves are given by $(dC/dx)^S$, and if plotted against t, produce the cottrell relation.

*It would be a mistake to believe that the error function, in this context, has any relation to errors.

A the profile is nearly linear, if the time and thickness of the cell are such that the function falls within this region, we speak of *thin-layer cell*. In this case the diffusion transport involves the entire volume of the cell, and the mathematical questions describing the relation between current and voltage are relatively simple.

Region A corresponds to arguments of the error function less than unity, or:

$$x_d < 2\sqrt{Dt} \tag{12.2}$$

Region B corresponds to a more complicated situation and is of little analytical interest. Region C, corresponds to arguments larger than about 2; this can be described by the condition:

$$x_d > 4\sqrt{Dt} \tag{12.3}$$

Beyond this point, the solution is practically untouched by diffusion, and the behaviour can be approximated by semi-infinite treatment. This discussion above is only qualitative in nature, since the presence of the outer boundary influences the concentration profile, due to the zero flux of matter at the outer wall.

Equation (12.1) depends on time, so that for a long enough period, any cell will behave like a thin-layer cell, and correspondingly, for short engouh times, any cell will obey semi-infinite mathematics. As a rule, the electrodes must be separated by less than 0.1 mm for thin-layer operation to prevail.

There are many ways of implementing such thin electrolyte layers, a few of which are shown in Figure 12.3. In cases *a* and *b*, the current must travel the length of a slice of solution, and resistance effects may become important. In *c*, this is eliminated by the use of a porous materials in contact with the electrolyte. This latter case is more properly discussed in the next section, as membrane electrode.

The fundamental characteristic of thin-layer operation is that the whole cell is essentially in equilibrium with the electrode. In other words, the quantities C^* and C^S are practically equal at all times. For this to be true, the cell must be operated in region *A* of Figure 12-2. For usuall cells, this requires experimental durations longer than about 1 s. Consequently, any potentiostatic operation, such as LSV, can be utilized if the scar is sufficiently slow.

It follows that the concentration ratio in the entire cell can be approximated by:

$$\frac{C_{OX}}{C_{RED}} = \exp\left\{\frac{nF}{RT} = (E - E^{o'})\right\} = \Theta \qquad (12.4)$$

which is a form of the Nernst equation. In this closed system, unlike LSV, $C_{ox} + C_{red} = C_{total}$. and thus:

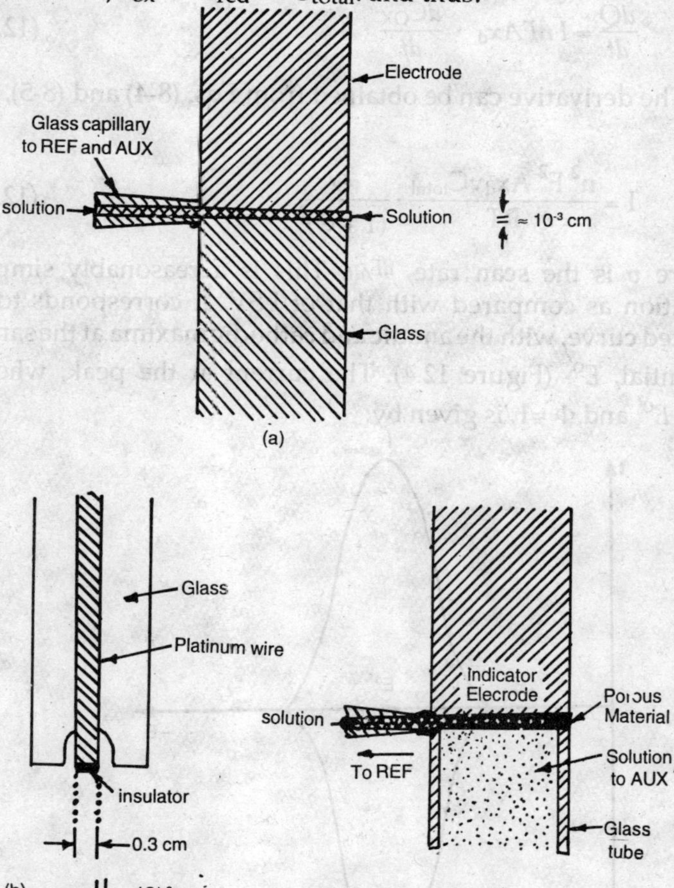

Fig. 12.3. Typical thin-layer cells. (*a*) Two parallel plates, one or both of which are electrodes; the plates can be positioned by micrometer mechanism. (*b*) Cylindrical cavity probe immersed in the solution. (*c*) Absorbent cell.

$$C_{OX} = C_{total}\left(1 - \frac{1}{1+\Theta}\right) \tag{12.5}$$

as the reader can easily ascertain. It is now possible to calculate the current as a function of time, since it is related to the chemical change by the factor nF. The total amount of the substance is Ax_dC, where A is the area and x_d the thickness. This corresponds to $Q = nFAx_dC$ coulombs, and the current is given by the derivative:

$$\frac{dQ}{dt} = I\, nFAx_d \cdot \frac{dC_{OX}}{dt} \tag{12.6}$$

The derivative can be obtained from Eqs. (8-4) and (8-5), to give.

$$I = \frac{n^2 F^2 Ax_d v C_{total}}{RT} \cdot \frac{\Theta}{(1+\Theta)^2} \tag{12.7}$$

where v is the scan rate, dE/dt. This is a reasonably simple equation as compared with that of LSV. It corresponds to a peaked curve, with the anodic and cathodic maxima at the same potential, $E^{o'}$ (Figure 12.4). The current at the peak, where $E = E^{o'}$ and $\Phi = 1$, is given by:

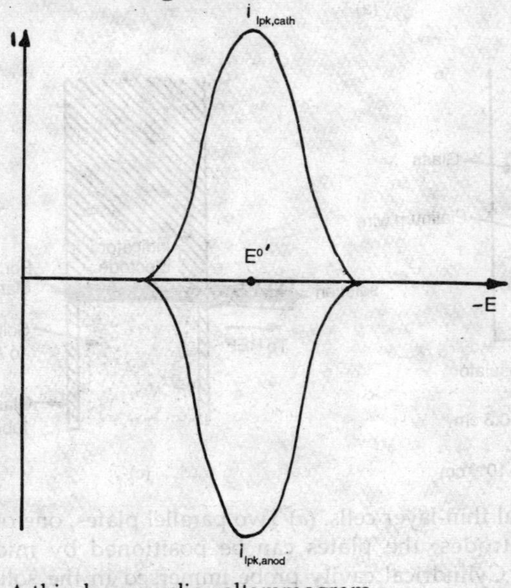

Fig. 12.4. A thin-layer cyclic voltammogram.

$$i_{pk} = \frac{n^2F^2 Ax_d v\, C_{total}}{4RT} \tag{12.8}$$

The behaviour is different from conventional LSV, where there is voltage shift between the anodic and cathodic peaks. The appearance of a shift in thin-layer operation would indicate that not all the solution is in equilibrium with the electrode. This can occur if the sweep rate is too fast (greater than about 10 mV/s), or in case of sluggish kinetics. This permits the use of thin-layer cells for kinetic measurement. For example, the transfer coefficient can easily be obtained from the peak heights :

$$\alpha = \frac{i_{pk}(cath)}{i_{pk}\,(cath)-i_{pk}\,(anod)} \tag{12.9}$$

Correspondingly siple relationships exist for the heterogeneous rate constant, which causes peak separation of about 120 mV per order of magnitude of k^o.

Galvanostatic operation of thin-layer cells in also of interest. If a constant current is applied, a given species will be electrolyzed in the *transition time:*

$$\tau = \frac{nF}{I} \cdot Ax_dC \tag{12.10}$$

where the product Ax_dC gives the number of moles present. In practice, the time τ is somewhat shorter than given by this eqution, since as the electrolysis proceeds, at some point there will be an insufficient amount of active substance to support the current, and the electrolysis of some other species will take place. The effect is more pronounced at higher currents and with thicker cells.

Coulometry, while particularly easy to apply in thin-layer cells, exhibits a sensitivity of only about 10^{-5} *M*. Note, however, that the sensitivity in total number of moles is outstanding, due to the small volume involved (typically 10 μ L or less), which means that only a fraction of a nanomole is reacted.

Even more advantageous is the use of pulse polarography, which the detection limit is about 10^{-7}M, using the same volume coresponding to only a micomole of reagant.

Many electrochemical methods involving semi-infinite diffusion can be adapted to thin-layer operation, often resulting in simpler mathematical modeling. The exhaustive nature of the process and the small volume of solution cause the ratio of active species to adsorbed impurity to be advantageous. This permits the study of monolayers and adsorption effects with much better experimental control than available with convention approaches.

Electrochemistry with Immobilized Reagents

The next logical step beyond thin-layer cells involves actually binding the active species to the surface of the electrode, which is tantamount to using a very thin cell. The process of immobilization can be affected by adsorption, by chemisorption, or by coating the electrode with a substance chemically bound to it. The amount of substance available is given by the surface excess, Γ, previously defined. Normally the electrode is loaded with regent in one solution and transferred to another for use. Under these conditions the surface excess represents the totality of substance involved in the redox process, which for an area A becomes the product AΓ. The concept of concentration is no longer valid in the conventional sense, and one must substitute the activity associated with a given surface excess.

$$\alpha_{OX} = \gamma_{OX}\Gamma_{OX} \tag{12.11}$$

and

$$\alpha_{RED} = \gamma_{RED}\Gamma_{RED} \tag{12.12}$$

where the quantities γ_{ox} and γ_{red} are proportionality constants with the units of cm^{-1}.

The behaviour of such immobilized substances depends on the status of both OX and RED. If only one of them is strongly held, the system follows a mixture of bounded and unbounded diffusion kinetics; if, however, both species are immobilized, the behaviour is quite similar to that of a thin-layer cell. For LSV the current is given in this latter case by:

$$I = 4_{ipk} \cdot \frac{\Theta}{(1 + \Theta)^2} \tag{12.13}$$

This should be compared with Eqs. (12.7) and (12.8).

The constant I_{pk} represents the equal heights of the anodic and cathodic peaks, which occur at the same potential for reversible reactions. The behaviour immobilized reagents has been amply documented both for adsorbed species and for chemically bound materials, the so-called *chemically modified electrodes.* Shifts in the potential of the peaks occur if $\gamma_{ox} \neq \gamma_{red}$, and also if the electrontransfer reaction is slow, as seen in thin-layer cells.

A somewhat different dimension is provided by the use of surface modifiers to alter the catalytic properties of an electrode surface. Thus if a reversibly reducible species is chemically bound in the form of a polymer layer, the electron transfer can occur in two steps:

$$(POL)_{OX} + e^- \rightarrow (POL)_{RED} \qquad (12.14)$$

$$(POL)_{RED} + OX \rightarrow (POL)_{OX} + RED \qquad (12.15)$$

so that the net reaction, $OX + e^- \rightarrow RED$, is controlled by the rate of reaction Eq. (12.14). The entity POL could be, for example, poly-(p-nitrostyrene) or polyvinylferrocene. The choices of active species and the modes of anchoring to the substrate are vast. Consequently one can foresee a great future in such electrodes, both in analysis and in preparative applications.

Membrane Electrodes

We Turn now to the non-bounded case of diffusion control. This occurs most commonly when the electrode is in contact with a membrane in which the diffusion in slower than in the unrestricted solution. In this case, the concentration profile will be similar to that indicated in Figure 12.5, established after a short initial transient.

The membrane can be considered as a porous material filled with the solution (in other words, there is only mechanical hindrance). The diffusion coefficient can be assumed to be basically unchanged by the presence of the membrane, and only modified by the fact the transport cannot occur in a straight path:

$$D_{effective} = D/q \qquad (12.16)$$

where q is a dimensionless number, the *tortuosity factor.* In addition, the electrode area must be corrected by a factor, ε, the porosity of the membrane, defined as the ratio of the area open to transport to the total area

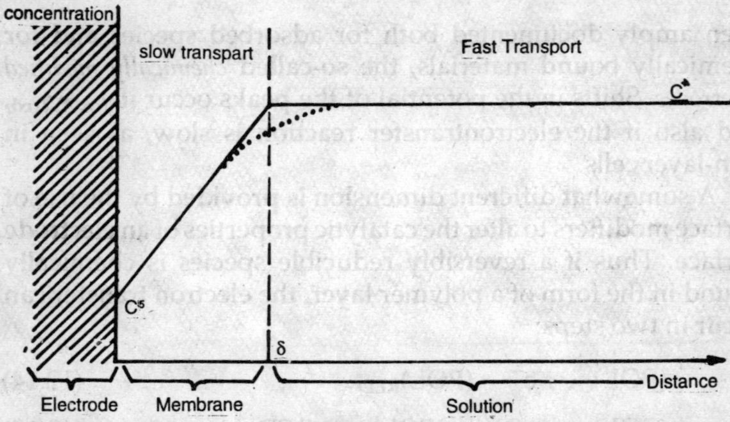

Fig. 12.5. The concentration profile in a membrane electrode. Full line: simple model; dotted line: actual profile.

The response of a membrane electrode of thickness δ, changes in several stages following the application of a potential step. Initially, it is time-dependent, and semi-infinite diffusion prevails. After a period of

$$t_1 = 0.2\frac{\delta^2}{D/q} \tag{12.17}$$

the process converts to a finite diffusion transient. At still times, this gives place to steady-state diffusion, which occurs at about

$$t_2 = 2\frac{\delta^2}{D/q} \tag{12.18}$$

It follows that one can operate a membrane electrode under semi-infinite or under thin-layer nonbounded diffusion depending on the time scale and the membrane thickness.

For long experiments, a steady-state currrent is attained, given by:

$$I = nF\ \frac{A\varepsilon}{\delta}\frac{D}{q}C^* \tag{12.19}$$

This relation can serve to measure the concentration C^*, and it can be employed as a convenient approach to the determination of the diffusion coefficient by comparison with a standard.

Membrane electrodes tend to be slow in their response. Thus for $\delta = 0.01$ cm, $D = 10^{-5}$ cm^2 s^{-1} and q = 10, a steady state is attained only after about 200 seconds. The presence of the membrane has a favourable isolating effect in solutions with solid suspensions, such as many biological fluids, where membrane electrodes have found some use. An example of *Clark oxygen sensor*, which employs a gas-permeable membrane. The electrode is highly selective, responding only to oxygen.

Ultramicroelectrodes

Another special case of diffusion occurs when the area of the electrode is very small ($<10^{-4}$ cm^2). Taking into account that the diffusion layer reaches this dimension after less than a millisecond, it can be seen that, as a result, the transport of active species is relatively constant. A linear sweep voltammogram generates a wave rather than a peak, as seen in Figure 12.6.

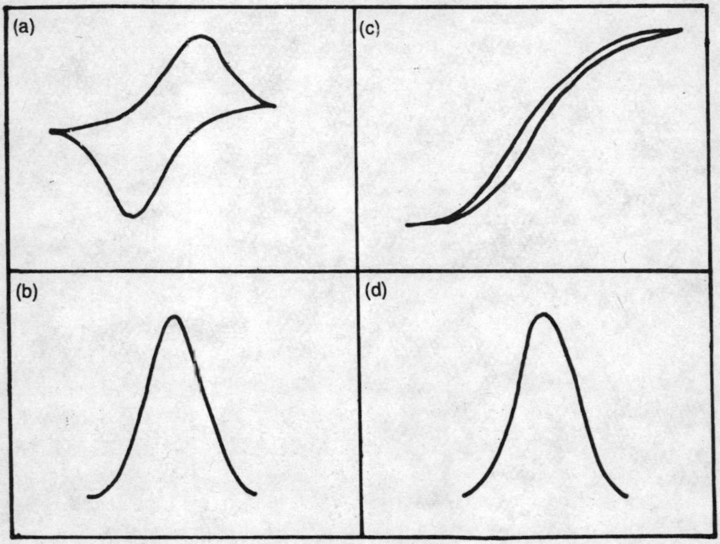

Fig. 12.6. Response of an ultramicro electrode to linear sweep voltammetry (*c*) and to DPP (*d*), compared with a conventional electrode in the same solution (*a* and *b*).